A BOOK
OF NOISES

The Book of Barely Imagined Beings:
A 21st-Century Bestiary

A New Map of Wonders:
A Journey in Search of Modern Marvels

CASPAR HENDERSON

A BOOK
OF NOISES

NOTES ON THE AURACULOUS

The University of Chicago Press

The University of Chicago Press, Chicago 60637
© 2023 by Caspar Henderson
All rights reserved. No part of this book may be used or reproduced in any
manner whatsoever without written permission, except in the case of brief
quotations in critical articles and reviews. For more information, contact the
University of Chicago Press, 1427 E. 60th St., Chicago, IL 60637.
The moral right of the author has been asserted.
Published 2023
Printed in the United States of America

32 31 30 29 28 27 26 25 24 23   1 2 3 4 5

ISBN-13: 978-0-226-82323-2 (cloth)
ISBN-13: 978-0-226-82324-9 (e-book)
DOI: https://doi.org/10.7208/chicago/9780226823249.001.0001

Originally published in English by Granta Publications as *A Book of Noises:
Notes on the Auraculous* copyright © Caspar Henderson, 2023
First published in Great Britain by Granta Books, 2023

Library of Congress Control Number: 2022061279

♾ This paper meets the requirements of ANSI/NISO z39.48-1992
(Permanence of Paper).

Music floats on wind
like driftwood on waves.

In the ocean depths,
pearls shine, lending their beauty
but never touching the surface.

We hear their dazzling echo.

Rumi
(Translated by Haleh Liza Gafori)

FIRST BREATH

A creation story from the
Guarani people, Paraguay.

Tupã, the first father of the Guarani people, stood up in the middle of the darkness and, inspired by the reflections of his own heart, created the flames and the thin fog, the beginning of a song.

While he still felt inspired, he created love, but he had no one to give it to. He created language, but no one could hear him speak it.

So Tupã recommended the gods to build the world and take care of the fire, fog, rain and wind. And he handed them the music with the words of the sacred hymn, so they could give life to the woman and man. Now the world would not be in silence at last.

So love became communion, and language took over life, and the first father redeemed his solitude in the company of the man and the woman who sing, 'we are walking this land. We are walking this shiny and beautiful land.'

Contents

Introduction

This book began in wonder. I was visiting a nature reserve on the Norfolk coast to see knots – short, stocky birds which migrate to Britain in large numbers each winter from their Arctic breeding ground. In flight, knots dash and turn together in such a way that they seem to appear and disappear as the light hits the bright and dark sides of their bodies. It was a beautiful sight, but what really struck me that day was the sound as thousands of them flew overhead. It was loud, but discernibly composed of lots of smaller noises – the fluttering of individual pairs of wings arriving at the ear fractions of a second apart with, perhaps, fractionally distinct timbres and microtones.

Afterwards, while I was working on a book titled *A New Map of Wonders*, I coined a term for this kind of experience. The sound of the knots was, I suggested, an instance not of the miraculous but the 'auraculous': a wonder for the ear – or, as the writer Robert Macfarlane helpfully glossed it, the 'ear marvellous'. And as I wrote, I realised how little I actually knew about sound in general and the various ways in which it shapes life. I decided to try to learn more, and this book is a result of that effort, though by no means its end.

The forty-eight entries in *A Book of Noises* are assigned to four categories. Three of the categories were conceived by the musician and soundscape ecologist Bernie Krause, and I have added one. 'Geophony', the first of Krause's categories, is for sounds of the Earth such as volcanoes, thunder, the Northern Lights and the rhythms of the planet that are not in themselves alive as we usually understand the term but which make life as we know it possible. The second, 'biophony', is for sounds of the living world. Here are explorations of some of the rhythms of the body, the nature of hearing, and the sound worlds of plants (yes, plants) and animals. The final category is 'anthropophony', a slightly awkward label for sounds associated with humanity. Under this heading I stumble and mumble about the origin and nature of language, music, harmony, the haiku of Bashō, strange musical instruments, the sounds of hell, climate change, noise pollution, healing and more.

My addition to Krause's categories is 'cosmophony' for sounds of the cosmos. This might seem odd because there is no sound in the vacuum of space. But resonance and sound play a fundamental role in the formation of all that is. In this category are entries on sounds that occur beyond the Earth, as well as those that people have imagined or projected on to and into space, from the music of the spheres to recent experiments in sonification that help listeners better conceive or imagine some of the things that are out there.

I have ordered the categories so that they start with this cosmic view, then focus in on this planet, then on the living world and then on human experience, but there is nothing to lose – and there may be something to gain – by jumping to whatever catches your eye first. Have a blast. Skip a whole category if you like. The entries in each category are just that – entrance points for longer journeys. There are many things I have scarcely touched on. To cite the Sufi philosopher 'Ayn al-Quḍāt in the conclusion to his book *The Essence of Reality*, 'there is no [more] time to elaborate, and I am fatigued. This is my excuse for any chapter whose topic I did not do justice to.'

Among the many places to learn more my recommendations include *Noise* by David Hendy, *Sounds Wild and Broken* by David George Haskell, *An Immense World* by Ed Yong, *The Sounds of Life* by Karen Bakker and *The Musical Human* by Michael Spitzer. I have also greatly enjoyed the warmth and range of *The Sound of Being Human: How Music Shapes Our Lives* by Jude Rogers, and delighted in the clips of wild sound, a different one each day, posted at earth.fm.

When I told people that I was working on a book about sound and noise I was quite often asked if a tree makes a sound when it falls in a forest but there is no one there to hear it? The short answer to this is yes: a trunk crashing down sends vibrations through the air whether or not anyone is listening. That's what sound is. But there is also a way in which the short answer is no, because sound as we usually think of it is an experience of a sentient being (and we tend to assume that trees and rocks are not sentient, or at least not in that way). If that's all you wanted to know, then you can put this book down now. But while these short answers may be true they are also unsatisfactory because there is, I think, often something else lurking behind the question concerning the listener's relation to the universe as represented for them by the forest. That unspoken (and perhaps unconscious) thought, I'd suggest, is something like, will the world *really* go on without me? It can be hard to get one's head around the idea that the world will continue without the awareness to which we as individuals so often cling. As Alexander von Humboldt wrote in 1800, 'This aspect of animated nature, in which man is nothing, has something in it strange and sad.'

Some sounds can be a kind of revelation to those who hear them, and sometimes the experience can be deeply unsettling. In Don DeLillo's novel *White Noise*, an air-raid siren in a residential neighbourhood that has been mute for a decade or more shrieks back into life, like a sonic monster, 'a territorial squawk from out of the Mesozoic. A parrot carnivore with a DC-9 wingspan.' And when, in his exploration of the world of those preparing for apocalypse, the writer Mark O'Connell visits a former US Air Force bunker that is being repurposed for end-of-the-world preppers, the sound of its great doors closing is like nothing he has ever heard: 'an overwhelmingly loud and deep detonation, the obliteration of the possibility of any sound but itself'. In a poem by W. S. Merwin, a foghorn becomes a 'throat' that 'does not call to anything human / But to something men had forgotten / That stirs under fog'. And in Apichatpong Weerasethakul's film *Memoria*, an extremely loud noise heard only by the protagonist foretells (and maybe causes) a descent, or possibly an ascent, into a strange dimension of existence – or annihilation.

But revelations in sound can also be comforting and life-expanding, bringing reassurance and beauty in the wide view. This is expressed in comic form by Roald Dahl's Big Friendly Giant, who 'is hearing the little ants chittering to each other as they scuddle around in the soil [and] is sometimes hearing faraway music coming from the stars in the sky'. It takes a mysterious, transcendental form in Jorge Luis Borges's short story 'The Aleph', where the faithful who gather at the great mosque of Amr in Cairo know that the hum of the entire universe can be heard by placing one's ear against one of the stone pillars in its central courtyard. The physician and essayist Lewis Thomas took pleasure in imagining all the non-human sounds of the Earth together: 'If we could listen to them all at once, fully orchestrated, in their immense ensemble,' he writes in 'The Music of This Sphere', 'we might become aware of the counterpoint, the balance of tones and timbres and harmonics, the sonorities.' And in one of his 'Love Letters to the Earth', the Zen monk Thích Nhất Hạnh writes that 'humanity has great composers, but how can our music compare to your celestial harmony with the sun and planets – or to the sound of the rising tide?'.

We live in times in which more is being destroyed than is being created. (Extinction rates of non-human forms of life, for example, are much higher now than at any time in Earth's history, including during mass extinction events millions of years ago.) 'Modernity stands at risk of no longer hearing the world and, for this very reason, losing its sense of itself,' writes the sociologist Hartmut Rosa. 'Our greatest fear should perhaps be that we have forgotten how to listen to the living Earth,' adds the biologist David George Haskell, who documents a catastrophic loss of sonic diversity and richness worldwide. And it is precisely because of this that it has never been more important to pay attention. Building on pioneering work a generation ago by the composer R. Murray Schafer and others, ecologists today are increasingly recording 'soundscapes' on land and in the ocean over the seasons and years as a means of assessing the vibrancy and health of ecosystems. By enabling us to listen more carefully and deeply, new technology can help us to limit and even reverse some of the damage that has been done.

For me, writing this book has been part of an attempt to listen more deeply and hold on to a sense of aliveness, in which each moment is a validation of what came before and a preparation for what comes after. I hope that reading it may be part of one for you. 'The human brain,' says the Blackfoot philosopher Leroy Little Bear, 'is like a station on the radio dial; parked in one spot, it is deaf to all the other stations ... the animals, rocks and trees simultaneously broadcasting across the whole spectrum of sentience.' William Blake would agree. 'Man has closed himself up, till he sees all things through narrow chinks of his cavern,' he wrote; 'and yet even locked in the cavern we hear distant echoes.' Come with me, and listen to some of them.

Oh, and one more thing. The words 'sound' and 'noise' are used interchangeably in this book. This may strike some readers for whom they have different associations or meanings as odd. 'Noise' is often associated with chaotic or unwanted sound. We tend to talk about noise pollution, for instance, rather than sound pollution. Noise is also a technical term in information theory, where it refers to random fluctuations in data that get in the way of a signal, and this usage has been borrowed by the psychologist Daniel Kahneman and others to describe chance variability in human decision-making. But 'noise' doesn't have to be a negative. In evolutionary and developmental biology, noise in a technical sense makes variation and hence innovation possible. And in both everyday speech and the poetic imagination it can speak of wonders. 'The isle is full of noises,' rejoices Caliban in *The Tempest*; 'Sounds, and sweet airs, that give delight and hurt not.' Further, the word 'sound' doesn't always have positive associations. As Macbeth begins to disintegrate he comes to see life as 'full of sound and fury, / Signifying nothing'. If Shakespeare was happy to use the words interchangeably, then maybe we can be too.

COSMOPHONY

Sounds of Space

First Sounds

For the first two to three hundred thousand years after the Big Bang the rapidly expanding universe reverberated as if filled with countless cosmic bells.

Sound is a pressure wave in a medium, and the denser the medium the faster it travels. The universe in these first millennia was so dense that it trapped light, but sound was able to pass through it freely at massively faster speeds than it does through the atmosphere on Earth today.

As everything cooled and atoms formed, the universe became transparent and light was able to travel too. Sound had concentrated matter along its wave fronts, and then, as the universe continued to expand, the resonance moved out in concentric waves, like the ripples on the surface of a pond after a handful of gravel has been thrown into it.

The wave peaks became foci for what later became galaxies. The universe we see is an echo of those early years, and the waves help us measure the size of the universe. The last chimes of the Big Bang get quieter and deeper as the universe expands.

But if, as some cosmologists maintain, our universe is just one in an infinite series, the cosmic bells have sounded many times before our universe began and will do so again after it ends.

Resonance (1)

Sound is governed by resonance, a phenomenon that shapes reality at every level – determining the existence of subatomic particles, the process that creates the atoms of life, the orbits of moons, and the run of the tides.

Resonance, explains the physicist Ben Brubaker, occurs when an object is subject to an oscillating force that is close to one of its 'natural', or resonant, frequencies. A simple example is a playground swing. Push on it at the right time and the swing will go higher – hopefully eliciting whoops of delight from a child on the swing. But however hard you push, the swing, which is in effect a pendulum, will resist variation from its natural frequency.

The understanding of resonance as integral to the cosmos owes a lot to Erwin Schrödinger. In 1925, ten years before he formulated a thought experiment to illustrate the conundrum of quantum superposition, in which a cat in a box is somehow both alive and dead at the same time, Schrödinger derived an equation to describe the behaviour of the hydrogen atom to which the solutions are waves oscillating at a set of natural frequencies. This equation is very like those that describe the acoustics of a musical instrument.

In the following decades, astrophysicists determined that

resonant transitions are also critical to the transmutation of one type of atomic nucleus into another in the super-hot and dense core of a star that is running out of fuel and collapsing in on itself. In one such nuclear resonance, three helium nuclei fuse into the nucleus of one atom of carbon. Without this musical alchemy, life would not exist.

Quantum field theory, which builds on the work of Schrödinger's contemporary Paul Dirac and others, holds that the universe's most basic entities are fields (which one may picture in general terms as something like what is revealed in the distribution of iron filings around a magnet). The elementary particles that constitute everything we know are actually local, resonant vibrations in these fields. It is by studying traces of these resonances that the existence of fundamental particles such as the Higgs Boson and the top quark has been confirmed.

William Blake asked his readers to see the world in a grain of sand. It might be no less fanciful to find it in a wave of the sea – or a child on a swing on an echoing green.

Sound in Space

It's not just the view from a balloon that can be amazing. The sounds can be too. When there is no wind sound travels upwards as easily as it does horizontally (and perhaps more so because it bounces up off the ground below), and reaches your ears with total clarity. Birdsong in a wood, the bark of a dog, the slam of a car door – I have heard all these to be pin-sharp when passing a few hundred feet or metres overhead. On a daring flight in 1836 from London across the English Channel and beyond, the pioneering balloonist Charles Green and his companions found themselves flying at night over Liège, at that time one of Europe's major industrial centres, and were overwhelmed by the thunderous machine noise below. 'There was,' records the historian Richard Holmes, 'disembodied shouting, coughing, swearing, metallic banging and sometimes, weirdly, sharp echoing bursts of laughter.' From a balloon, the world beneath becomes not just a panorama but a panacousticon, in which everything is audible.

Float higher than a few hundred feet, however, and most sounds on the ground start to become too faint for the human ear. At a height of twenty-one kilometres, or thirteen miles, the current world altitude record for manned balloon flight, you'd need sophisticated microphones to detect anything. The deep blue air is not endless; by the time you're eighty kilometres (fifty miles) up it is so thin that the only sounds that can pass through it are at frequencies typically below the range of human hearing such as those from earthquakes. Above the Kármán line, at 100 kilometres (sixty-two miles), begins a vast expanse with almost no sound at all – except, perhaps, for the occasional billionaire shouting 'Whee!'.

Sound can travel wherever matter is sufficiently concentrated, so it is present in and on stars, planets and other concentrations of atoms in space just as it is on the isle of noises that is the Earth. Convection currents on the surface of the Sun create sound waves that would be as loud as a jackhammer on Earth if air as thick as our lower atmosphere extended all the way there. The effect would be a dull roar, rather like standing next to Niagara Falls but about twice as loud. Sound waves also bounce around inside the Sun, and astro-seismologists study them to 'see' vast rivers of material flowing around deep in the interior.

At a much larger scale, sound waves reverberate inside super-bubbles – cavities hundreds of light years across which are made by the stellar winds and supernova explosions of stars eighty to a hundred times the mass of the Sun. The sounds are very deep, though not as deep as those emitted by black holes such as the one at the centre of the Perseus cluster, which oscillates once every 10 million years or so. In a 'remix' published in 2022, the Perseus waves are scaled up by fifty-seven and fifty-eight octaves. The result is rather like the groans of a ghost in a bottomless well.

There is sound within and on planets and moons in our solar system too. Mercury has no atmosphere to speak of and so is silent above ground but, pulled and yanked by the Sun, the planet is subject to seismic activity. If you were to put your ear – or more likely some seismological equipment – to its surface this would be clearly detectable. Seismometers placed on Earth's moon enable researchers to measure shudders and groans that are mainly caused by meteor impacts, and the squeezing and stretching of its interior by the tidal pull of the Earth. Quakes on Mars, which appear to be produced by the planet cooling and contracting, enable researchers to map the planet's interior. Some also hope to install devices to measure seismic wave propagation on Jupiter's moons Ganymede and Europa, and Saturn's moon Enceladus one day.

In contrast to Mercury, Venus has an atmosphere that conducts sound rather well. At ground level this 'air' is a super-critical fluid of carbon dioxide more than ninety times as dense as the atmosphere on Earth. If, as seems likely, there is thunder that accompanies the lightning that tears through the Venusian sky, it would reach one's ears quickly but could be rather muffled, and at higher frequencies.

In 2012 researchers modelled the effects that the different atmospheres, pressures and temperatures on Venus, Mars and Saturn's moon Titan would have on the human voice and other sounds. Setting aside the detail that on Venus a human being would be crushed and burned up almost instantaneously, the pitch of your voice would become much deeper as the vocal cords vibrated more slowly in the gassy soup of its atmosphere. However, because the speed of sound is much faster than it is on Earth, our brain would judge the speaker to be further away. Humans on Venus, the researchers concluded, would sound like bass Smurfs.

On the surface of Mars the atmosphere is about a hundredth as dense as it is at sea level on Earth – or about the same as it is at thirty-five kilometres (twenty-two miles) above our heads. Martian air, which is mostly carbon dioxide, is extremely cold, and this reduces the speed of sound and so would lower the pitch of a voice. On the other hand, the low density of the air would raise the pitch, and it is thought that the two factors would roughly balance out so that overall we'd sound much the same on Mars as on Earth, except that our voices would be very faint.

The thinness of Martian air means that even the great storms that sometimes blow would feel like zephyrs and gentle spirits of the air to a human standing on the planet. The only ambient sound one might hear would be dust and sand bouncing off the faceplate of one's space helmet. It has become possible, however, to listen remotely to the actual sound of Martian wind blowing at much lower speed. Recordings made by the Perseverance Rover and beamed back to Earth in March 2021 reveal it to sound pretty much as you might imagine: an empty gusting in one of the most desolate places imaginable, where no water has flowed for billions of years. The following month Perseverance captured the sound of its tiny helicopter drone, Ingenuity. It sounds pretty much like a drone on Earth, though marginally deeper.

Our understanding of what sound would be like in the atmospheres of planets and moons further out in the solar system is more speculative. Jupiter's atmosphere is mostly made of hydrogen and helium, which would raise the pitch of a human voice. The cloud decks of this giant planet are frequently shaken by lightning vastly more powerful than any on Earth, and the resulting thunder may carom across distances many times greater than the diameter of the Earth. On the surface of Titan, a moon unique in the solar system for having a thick atmosphere, liquid methane falls as rain and may flow across the rocky surface much as water does on Earth. Sand dunes similar in appearance to those in the deserts of Namibia may sometimes 'sing' in the wind. Here, where the average temperature is minus 182.5 °C (minus 296.5 °F), the sounds will probably be deeper than we can easily imagine.

Space, *The Hitchhiker's Guide to the Galaxy* helpfully explains, is big. The corollary is that 'tiny' does not begin to describe how mind-squishingly small the Earth is by comparison. But the vast cosmos, which started with sound, is now silent for the most part. The idea of such an unimaginably large sonic abyss can induce a sense of existential vertigo. That prospect, however, need not terrify. The musician Jordi Savall says that he prefers to record between about two and four in the morning, when 'you can feel the deepness of the universe because the silence is so enormous'. And so it is that awareness of the void can be a gateway to living with more delight in what one can experience, co-create and share of this world and whatever exists beyond.

Music of the Spheres (1)

On a gentle summer night when the Moon, Venus, Mars, Jupiter and Saturn hang like lanterns in the sky, or when countless stars are sparkling, it can be hard not to feel that beyond the silence there is a kind of music in the air. I would be reluctant to compare it to any actual melody, but a sequence of tracks titled 'Sublunar' on Max Richter's album *Sleep* is roughly there.

I know that the music I imagine when I gaze into the night sky is not real – that there is no sound which passes between the stars and planets and me. I know too that what I feel has been shaped by specific cultural traditions. There is the ecstasy of Rumi: 'We have fallen into the place where everything is music.' There's Dante's *Paradiso*, filled with harmonies in comparison to which even the sweetest sounds on Earth are storm and fury. And there are the lovers imagined by Shakespeare for whom 'There's not the smallest orb [in heaven] which thou behold'st / But in his motion like an angel sings'. But I still wonder if what I feel is, at least in part, an expression of something deeper than any accident of culture.

For at least two thousand years many people in Europe and beyond believed that the movements of the heavenly bodies created a universal harmony – a music of the spheres – that could be understood and appreciated in terms of harmonic and mathematical relationships which linked human life to a divine order. The idea is said to have formalised in the sixth century BCE with Pythagoras. This philosopher and mystic, about whom little is known for sure, was supposedly the first to notice that intervals between notes that sound harmonious on Earth can be described by simple ratios of size and distance. Pythagoras suggested that, on the same principle, the heavens are a kind of musical instrument, with celestial bodies each producing their own notes in proportion to their different orbits around a common centre.

For Pythagoras and his followers, the essence of everything was number. They believed that the universe was sustained by harmony in a perfect, eternal order, and that the music of the spheres shaped life on Earth. Making music, in imitation of the heavens, was an essential part of their practice and was intended to rouse, calm and purify the soul.

The Pythagorean school was influential across the Greek world and beyond. The correspondence it had identified between physical dimensions and sound was, arguably, one of the first laws of physics. Plato, who lived some three generations after Pythagoras, described astronomy and music as twin sciences, both accessible to humans through their senses. 'As the eyes . . . seem formed for studying astronomy,' he wrote, 'so do the ears seem formed for harmonious motions.' But not everyone agreed. Aristotle doubted the existence of celestial music, arguing that if it were real it would be so loud that it would shatter the Earth. The Roman statesman and philosopher Cicero suggested a solution (or a fudge): the sounds were real but, just as our eyes are not equipped to look at the Sun, so our ears were not capable of hearing them and the other heavenly bodies.

The oldest surviving attempt to assign note values to the orbits of the spheres is in *The Manual of Harmonics* by Nicomachus of Gerasa, a mathematician born in 60 CE in what is now Jordan. Pythagoras had supposedly calculated the distance from the Earth to the Moon to be about 79 million paces, and made this the celestial equivalent of a whole musical tone. Nicomachus proposed a seven-note sequence starting on a D, which he assigned to the Moon as the fastest-moving of the heavenly bodies, and descending with the Sun and planets through the natural notes, except for B, which is flattened. The whole made up a D natural minor scale. Other philosophers and musicians proposed a sequence covering two octaves in which the fixed notes were either a perfect fourth or a tone apart. This made for a more harmonious chord because the notes were not all scrunched together.

Around 510 CE the Roman philosopher Boethius tried to establish a systematic foundation for the understanding of all forms of music. There were, he said, three kinds: *musica mundana*, the music of the spheres; *musica humana*, the music of the human body and soul; and *musica instrumentalis,* the music that is played on instruments or sung and that we can readily hear and feel. The cosmic sounds of *musica mundana* were real but inaudible. Even so, nature resonated with heavenly song, which shaped life on Earth and caused the change of seasons. Like earlier authors and theorists, Boethius marvelled at the power of the music we can hear – *musica instrumentalis* – to arouse strong emotion. 'Music is so united with us that we cannot be free from it even if we so desire.' But he also warned that, depending on the form it took, this kind of music could either ennoble or degrade humanity.

The music of the spheres received renewed attention in Renaissance Italy. In his 1496 book *The Practice of Music*, Franchino Gaffurio argued that, just as astrology explained how the position of planets shaped human behaviour, so music linked the heavens and the soul. The frontispiece of the book shows a cosmic serpent with the Earth at its head and the planets and muses strung along its body, but instead of assigning each planet a single note Gaffurio attributes each a whole different scale or mode. His innovation reflected changes in musical style, especially a transition from largely single melodic lines to polyphony – the harmonisation of many voices together – in which his contemporaries, including his friends Leonardo da Vinci and Josquin des Prez, were discovering new ranges of moods and feelings. According to Gaffurio, each planet sings according to its mode and their individual melodies mingle in an ever-changing whole that mirrors events on Earth.

Boethius's great work, printed for the first time in 1491, almost nine hundred years after it was written, had fascinated Gaffurio and his contemporaries, but it also kicked up challenges. Following Pythagoras, ancient musical theory had held that the only truly consonant musical intervals were the octave and the fifth, and the tuning for the twelve half-tones of a scale was built by 'stacking' notes in fifths. The problem was that constructing a scale in this way did not produce a perfect octave, but landed about a quarter-tone off, producing a dissonance known as the Pythagorean comma: a glitch in the mechanism of heavenly harmony.

By the late fifteenth century European music was using intervals such as the third and the sixth to striking effect, challenging the Pythagorean ideal of the fifth and fourth as the only pure harmonies, and presenting problems for Pythagorean tuning, especially when moving from one key to another. A solution was found in the writings of Aristoxenus, an ancient Greek critic of Pythagorean musical theory whose *Elements of Harmony*, translated into Latin for the first time in 1564, suggested that the octave should be divided into twelve equal steps. This challenged received ideas about musical intervals and hinted at flaws in the unified cosmological and musical theory that would undermine it a few decades later.

The lutenist and composer Vincenzo Galilei championed Aristoxenus's system, and may have seen in it a corollary for the 'new' Sun-centred model of the solar system suggested by Copernicus (which was actually a revival of an idea first advanced by Aristarchus of Samos in the third century BCE). In his *Dialogue on Ancient and Modern Music* of 1580 Galilei does not mention heliocentrism, the heresy for which his son Galileo Galilei was to get into a lot of trouble in the 1630s, but it looks very much as if he had it in mind when he compared the notes in the octave to the planets in the night sky. 'Like the many lines drawn from the centre of a circle to the circumference which all gaze back at the centre,' he wrote, 'so every musical interval in the octave sees itself as if in a mirror, just as the planets do in the Sun.'

Ironically, the Pythagorean vision of a few elementary ratios at the heart of both cosmology and music finally came unstuck thanks to work undertaken in order to verify it. From an early age, Johannes Kepler, a contemporary of Vincenzo's son Galileo, had believed himself to be destined to understand the harmony of the universe, and in *The Harmony of the World*, published in 1619, he set out what he believed would be its definitive form.

Previously, Kepler had shown that the orbits of planets around the Sun were not circular, as Copernicus had supposed, but elliptical, and that the planets sped up and slowed down according to their distance from the Sun. Moving from data and observation to conjecture, Kepler assigned vocal ranges to each one. Mercury was the soprano, Earth and Venus the altos, Mars the tenor, and Jupiter and Saturn the basses. The variations in speed meant that their notes would change during the course of their orbits. The note from Mercury, which has the most elliptical path, varied the most. Venus has an almost circular orbit so its note hardly changed, but Earth's ranged between two notes a semitone apart. Kepler labelled these as Mi and Fa, and found in this minor interval what he considered an appropriately sad sound for a planet where, he felt, misery and famine held sway.

Kepler had hoped that the ensemble would produce a complex, ever-changing harmony. But he realised that the six lines would actually clash for most of the time as they slid between intervals in a way that was disturbingly alien for the music of his time. He also realised that the planets would never repeat their configurations: there would be no return to an original glorious harmony. These discoveries set his contemporaries and successors wondering whether an ever-changing polyphony of unheard sounds was really any use in understanding the heavens. By contrast, Kepler's laws of planetary motion, which could be expressed in mathematical terms without any reference to music, predicted the locations of planets in the past and the future with precision. The idea that music somehow ruled or explained the universe began to seem less plausible.

And yet there is a way in which the idea of celestial music did not die, but rather transformed and emerged in different forms to inspire musicians and others to dream of new kinds of music in this solar system and far beyond.

Music of the Spheres (2)

In 1977, some three and a half centuries after Johannes Kepler published his data on the movement of the planets in the solar system as *The Harmony of the World*, a jazz musician and a geologist turned it into an album. Willie Ruff and John Rodgers assigned each planet different notes according to the speed and shape of their orbits around the Sun and rendered the notes with an electronic synthesiser, layering them one by one to create a 'sound-picture' of the heavens.

Their album, also titled *The Harmony of the World*, begins with a thin, high warble, like a distant car alarm or a distressed Clanger muffled in a box. This is Mercury, sliding rapidly from an E8, the note just above the top of the range of a piano keyboard, down by a third to a C8 and back up again – the variation representing its pronouncedly elliptical orbit. Then comes Venus, sounding two octaves lower than Mercury at E6, and varying by only a quarter-tone over a three-second cycle in imitation of its almost circular orbit. Earth joins in somewhere from a sixth to a fifth below Venus on a G5, and wobbles up and down by about half a tone over a five-second cycle. Together, Venus and Earth create a dyad that changes from major-ish to minor-ish and back again. Mars enters on a C5 (an octave above middle C), and sweeps down three whole tones over the course of about ten seconds to an F4 ♯. Jupiter sounds a D at the bottom of the range of a piano and wobbles down to the B below and back up again. Saturn is a deep growl an octave below that. Ruff and Rodgers represent the orbits of the outer plants Neptune, Uranus and Pluto (which were unknown to Kepler) with drumbeats.

Here, near as dammit, is a true realisation of music of the spheres. And it is not a lovely thing. An account published in *The New York Times* on the music's first outing describes it, accurately, as 'a cacophony of tweedling, wailing, thumping and droning'. The newspaper reported that a six-year-old boy said it made him feel dizzy, while an adult man said it suggested motion sickness. Even Willie Ruff, its co-creator, conceded that it was 'very wearing' to listen to. And yet here is a kind of rebirth, or at least a milestone, on a path through sound to a better understanding of celestial mechanics.

It is often the case that nothing beats a good visualisation. Consider the 2015 film *To Scale: The Solar System*, which shows how you need an area more than eleven kilometres (seven miles) across for a model in which the Sun is about a metre and a half across and the Earth is the size of a marble. But seeing alone is sometimes not enough and, crude as it is by today's standards, Ruff and Rodgers's audio composition can help the listener to focus in more deeply – to sense the relative dimensions and movements of the planets in time as well as space. And their work was only a start. Thanks to human creativity and huge increases in computing power, the practice of sonification – turning data into sound – has flourished in the intervening decades.

The roots of this new era date back to the discovery in the 1930s that the Milky Way beams out radio waves. This was the birth of radio astronomy: the study of stars and other phenomena in space by monitoring the electromagnetic waves they emit at radio frequencies. Those waves can easily be turned into sound, just like radio waves on Earth, but for many years scientists saw little point in doing so, preferring rather to study them as numbers or graphic representations. Over time, however, this has changed.

In the early 2000s an astronomy student named Wanda Díaz-Merced was finding it increasingly difficult to study as diabetic retinopathy gradually destroyed her vision. One day a university friend played her a sound file depicting a huge ejection of matter and energy from the Sun. 'It was inspiring,' Díaz-Merced later told the Royal Society. 'I could hear the Sun in real-time, and when the sunburst finished, I could hear the galactic background.' She realised that not only could the light be transformed into sound, but that with sonification she might be able to continue her work; and using sound Díaz-Merced went on to research the light emitted by gamma-ray bursts, the most energetic events in the universe. She also showed that sonification could help sighted astronomers detect subtle signals indicating the presence of a black hole that they had missed in visual examination of data.

In 2020 astronomers working together with musicians showed that sonification can create beautiful as well as accurate representations of the night sky. Compiling data from the Chandra X-Ray Observatory, the Hubble Space Telescope and the Spitzer Space Telescope, Kimberly Arcand, Matt Russo and Andrew Santaguida created 'Sounds from Around the Milky Way'. Over the course of about a minute a bar sweeps across an image of the galactic centre. As it does so, various sounds represent the varying brightness of the stars and other sources of different forms of electromagnetic radiation. X-rays produce gentle chimes, infrared makes harp-like sounds, while visible light resembles the plucking of violins or cellos. The result is a gentle, ambient, aleatoric; rather as if irregular but recurring bubbles, eddies and currents of a heavenly stream have been rendered with musical instruments.

In 'Sounds from Around the Milky Way' and in other works such as 'Chandra Deep Field South' the team has created sonic wonder from the semi-chaotic distribution of energy, whether it be across the galactic centre or in the distribution of galaxies and black holes in deep space. But there are also circumstances in which celestial bodies move with almost perfect regularity with respect to each other. In what is known as orbital resonance, two or more bodies orbiting a common centre exert a regular, periodic gravitational influence on each other. Usually the bodies exchange momentum and shift orbits until the resonance no longer exists. To adapt Kafka on hope, there is harmony, but not for you. But every now and then a resonant system can be self-correcting and stable. And though Kepler did not know it, there are instances of such accord in the solar system. The moons Io, Europa and Ganymede, for example, orbit Jupiter in a perfect 1:2:4 ratio. This means that it takes Europa, the second-closest moon, twice as long to orbit Jupiter as it does Io, the closest, while Ganymede, the third-closest moon, takes four times as long as Io.

The 'resonance' in orbital resonance is one of gravity rather than sound waves, but it can easily be converted into the latter. Speeding a model of the Jovian moons up thousands of times so that they take seconds rather than hours to complete an orbit, a science-art outreach group called SYSTEM Sounds has rendered them as a tight, but quite funky, pattern of drumbeats. When the model is accelerated even further – to 250 million times actual speed – the rhythm becomes pitch. With Io travelling at twice the speed of Europa, and Ganymede travelling twice as fast as Europa, the three sound the same note across three octaves. (This is because a doubling of wavelength is perceived by humans as an octave.)

Orbital resonances have now been detected elsewhere in the galaxy. Between 2015 and 2017 astronomers discovered a resonant chain of seven Earth-sized planets circling TRAPPIST-1, a red dwarf star about forty light years from Earth. The whole system is small and tightly configured: TRAPPIST-1 itself is about the size of Jupiter and the planets going around it are all closer to it than Mercury is to our Sun. Going from innermost to outermost, the orbits are in ratios that approximate 8:5, 5:3, 3:2, 3:2, 4:3 and 3:2 respectively. These correspond to intervals of minor sixths, perfect fifths and a perfect fourth. If, in a model, the outermost (and therefore slowest and lowest) planet is sped up so that it sounds C3 (an octave below middle C), the others sound G3, C4, G4, D5, B5 and G6 – a cosmic C major 9 chord.

The harmony in the TRAPPIST-1 system is not perfect – the Cmaj9 is just a little out of tune – but it seems to have been good enough to keep the planets nudging each other in such a way that the ratios of their respective orbits have remained stable for at least 50 million years. Music – harmony – makes the whole endure.

In 2017 and 2018 citizen scientists identified another planetary system that comes close to harmonic perfection. The five inner planets of K2-138, a main sequence star nearly 600 light years from Earth, all orbit in almost exactly a 3:2 ratio with respect to their neighbours. When converted to sound the result is a stack of fifths. Not all of them are exactly in tune today, but modelling indicates that when, 2 billion years ago, the planets formed out of a disc of dust orbiting the infant star, they were. At its inception the K2-138 system made the perfect music of the spheres of which Pythagoras dreamed.

Discoveries like these may prove to be small glimpses of a vast new arena. It is less than thirty years since exoplanets (that is, planets orbiting other stars besides the Sun) were a theoretical possibility that had never actually been seen. The thousands that have now been identified are almost certainly only a tiny fraction of those in our galaxy, let alone those in the billions of other galaxies in the observable universe. There may be almost no end to the variations on a simple theme.

The idea of a music of the spheres as conceived in antiquity and celebrated in medieval and Renaissance Europe has enduring appeal. Johanna Beyer titled a pioneering 1938 electronic work *The Music of the Spheres*. Less remarkably, Coldplay gave this name to a 2021 album. The main theme from the 1977 film *Star Wars* by John Williams, which sits squarely in a tradition of big orchestral film scores and their nineteenth-century symphonic antecedents, is built with fifths that Pythagoras would have recognised immediately. So is the much more complex and demanding *Star-Child* by George Crumb, also from 1977. Jem Finer's *Longplayer*, which began in 1999 and will play without repetition for a thousand years, will eventually repeat its sequence of chimes in what the composer describes as akin to a system of planets where every movement is predetermined. In *Just Ancient Loops II* by Michael Harrison (2012), a cello plays musical intervals with reference to the orbits of Jupiter's four inner moons according to a tuning system that dates to classical antiquity.

Musicians and others have also responded to profound changes in human understanding of the cosmos that have unfolded since the time of Kepler and Galileo. In 'Supernova' (2017), Trevor Wishart converts the light spectrum of a giant stellar explosion and the spectra of the new elements it generates into sound. William Basinski's *On Time Out of Time* (2019) is composed with data from gravitational waves emitted during the merger of two black holes 1.3 billion years ago. In Max Richter's 'CP1919', which was recorded in 2020, pulsations and rhythms 'governed by the same ratios supposed by the ancient astronomers to describe the orbits of the planets' are deployed in homage to the first discovery, in 1967, of a pulsar – a compact star that emits regular bursts of radiation. (An earlier tribute to this discovery features on the cover artwork for Joy Division's 1979 album *Unknown Pleasures*.) These works – and many others, including perhaps *Stellar Regions* by John Coltrane (1967), *Asteroid 4179: Toutatis* by Kaija Saariaho (2005), *Promises* by

Pharoah Sanders (2021) and *Space 1.8* by Nina Sinephro (2021) –
explore the question of what a music commensurate with the
nature of the universe might sound like.

Four hundred years ago Kepler found to his dismay that the
planets he saw in the night sky did not move according to long-
established notions of perfection, and that there was dissonance
in the heavens. But what seemed like bad news to Kepler is actually good news for us. Had there not been slight irregularity in
the distribution of energy in the early universe, matter would
not have concentrated into stars and galaxies. And it's possible
that there would be no life on Earth but for what, according to
one hypothesis, was an epic bender by Jupiter early in the history of the solar system, when the giant planet spiralled millions
of miles inwards towards the Sun before tacking out again to its
current orbit – a manoeuvre that may have pushed water-rich
asteroids into the Earth, creating the ocean. If there had been no
lumps in the ancient cosmos there would be no 'Shape of You',
and if there had been no Jupiter wobble there might have been
no *Jupiter* symphony.

Music that is entirely regular and predictable verges on the
inhuman. There is beauty in imperfection. Error – not in the
sense of a mistake, but rather its Latin route *errare*: to wander
or go astray, and its Sanskrit cousin *arsati*: to flow – can be generative. There are limits to what the ear can process and the
mind comprehend but, amid the nothing that is not there and
the nothing that is, a new music of the spheres may pour forth
surprises and possibilities we hardly begin to imagine.

The Golden Record

Out beyond the edge of the solar system, two spacecraft are heading away from the Sun at more than ten miles per second. Onboard, they carry sounds of Earth, engrooved on old-style long-play records made of gold-plated copper and engineered to last more than a billion years.

NASA launched the craft, Voyager 1 and Voyager 2, in 1977 to study Jupiter and Saturn. Taking advantage of a favourable alignment of the outer planets, Voyager 2 also flew close by Uranus and Neptune, the only probe ever to have done so, and beamed back additional images and data. Meanwhile, Voyager 1, which had already flown out beyond the orbit of Neptune, turned its camera round, and in 1990 took a photograph in which the Earth appears as a single pixel – famously described by the astronomer Carl Sagan as a pale blue dot.

Sling-shotted outwards by the gravitational fields of the planets they passed, both craft will continue towards the stars indefinitely. In about 296,000 years Voyager 2 will pass within 4.6 light years of Sirius, the brightest star in the night sky. At the time of writing both Voyagers are still sending data home, but by the mid 2020s they will finally run out of power and fall silent. After that they will only 'speak' again in the astronomically unlikely event that an intelligent entity finds one or both and plays the records they carry. All other things shall change, but they remain the same, till heaven's changes have their course, and time hath lost his name.

The sounds on the records, chosen by a small group convened by Sagan, include surf on a beach, wind and thunder, the songs of birds and whales, greetings in fifty-five languages, and electrical signals from the brain of a woman in love. There is an electronic composition by Laurie Siegel based on Kepler's *The Harmony of the World*. There are also messages and images encoded on the records: printed messages from US President Carter and UN Secretary General Waldheim, photographs of everyday life tinged with an orangey-brown characteristic of the 1970s, and images of landscapes, plants, animals and the human body – although detailed depictions of genitals and the belly of a pregnant woman were not considered acceptable by NASA. A diagram on the cover of the records locates the Sun in relation to the different rhythmical beats of fourteen nearby pulsars – stars that emit rotating beams of electromagnetic radiation that sweep out across space like the signals from lighthouses.

One of the greatest treasures on the discs is proclaimed in words etched into them by hand: 'To the makers of music – all worlds, all times.' The twenty-seven tracks range from Bach to Chuck Berry, and are so varied in origin and kind that they escape a quick summary. One of the things that stands out in many of them, however, is the human singing voice in all its brilliance and directness. This is particularly so in tracks such as 'Tchakrulo', a three-part harmonisation for male voices from Georgia, and the Bulgarian 'Izlel je Delyo Hagdutin', in which pipes accompany a female voice that is both sweet and awesomely powerful.

Among the final four tracks are three which suggest that something of what is most essential in the human spirit is better expressed in music without words. The first of these is 'Flowing

Streams', a Chinese classic composed more than two thousand years ago. Performed by Guan Pinghu on an unaccompanied *guqin*, or seven-stringed zither, the music is meditative and in some respects quite simple. As well as depicting the river, which the ancients believed to be the blood of the world, 'Flowing Streams' is said to tell of a great friendship between its composer, Bo Ya, and a woodsman named Zhong Ziqi. It is said that when Ziqi died, Bo Ya broke the strings of his instrument and vowed never to play the piece again. Succeeding generations have continued to regard the music as a memorial to friendship as well as the beauty of life.

Ann Druyan, who was creative director for the Golden Record, identified 'Flowing Streams' with the help of a musicologist very near to the deadline for production and dispatch of the discs. Excitedly, she left a message on Sagan's answering machine, describing her discovery. The two had been having an affair, and when Sagan called back he asked her to marry him. She accepted. This romance then found its way into the Golden Record as it was electrical signals from Druyan's brain that featured on the recording of a young woman passionately in love. In 2017, the folk singer Jim Moray put this story at the centre of his song 'Sounds of Earth'.

The last track but one on the Golden Record is 'Dark Was the Night, Cold Was the Ground' by Blind Willie Johnson. Recorded in 1927, the sound is scratchy – the auditory equivalent of an old-time movie or a fading photograph. The guitar comes first: a bottleneck, metallic slide through blue notes in an introductory phrase before picking out the tune. Johnson begins to sing about half a minute in. He hums and aahs but never quite articulates the desolate words of the song ('Dark was the night, and cold the ground / On which the Lord was laid; / His sweat like drops of blood ran down; / In agony he prayed . . .'). But in a little over three minutes, the track helped define blues and slide guitar for generations to come. Pier Paolo

Pasolini used the recording in his 1964 film *The Gospel According to St Matthew*, and in 1984 Ry Cooder based the title track for *Paris, Texas* on what he called 'the most soulful, transcendent piece in all American music'.

The final track on the Golden Record is the cavatina from Beethoven's String Quartet No. 13. A cavatina is a short, simple song, and Beethoven has composed just that: a tender theme in two parts that he asks to be played *adagio molto espressivo* – slowly and very expressively. The simplicity of the piece is all the more striking given the context, for the thirteenth quartet is one of the longest and most ambitious of the late quartets, and Beethoven intended the cavatina to be played before the fiendishly complex and frenetic Grosse Fuge, or Great Fugue. It's a little like passing through a forest glade dappled in sunlight before ascending a mountain range.

People hear different things in the cavatina, observes the musicologist Philip Radcliffe. 'Some have felt it to be a profoundly tragic piece of music, while others have stressed its serenity, or its religious fervour.' Ann Druyan reports that when she first heard it she felt the cavatina captured a kind of 'human longing and even human hopefulness in the face of great sadness and great fear' that moved her profoundly. The Golden Record became her 'big chance to pay Beethoven back'. Later, Druyan discovered two things that amazed her. First, the composer had actually thought about the possibility of his music going into space, writing on the margin of one of his pieces: 'What will they think of my music on the star of Urania?' (a reference to the planet Uranus, which had been discovered when Beethoven was a boy). Second, on the manuscript for the quartet containing the cavatina he wrote *Sehnsucht* – the German word for longing. 'That affected me deeply,' says Druyan, 'because that was at the heart of the Voyager record: longing for peace and longing to make contact with the cosmos.'

The two copies of the Golden Record were shot into space

nearly fifty years ago. Today, an electronic device could store vastly more information – although it would be unlikely to endure for so long. There has been no end of talk as to which music and images would go on a new mission. The comedian Steve Martin joked that aliens had already been in touch to say 'send more Chuck Berry'. The composer Philip Glass has suggested Bach's cello suites. Bach, whose music already accounts for more than a tenth of all the music on the Golden Record, 'takes you by the hand ... and walks you into states of being that you didn't even know existed'. He also suggested music from Africa, throat singing, and flute-playing from South India. But the writer Mireille Juchau argues that a satisfactory set of choices could be beyond reach: 'The longer I consider it, the less possible it seems to capture the simultaneous unfolding of beauty and ruin in our present moment.'

GEOPHONY
Sounds of Earth

Rhythm (1) – Planet Waves

The earliest known depictions of Shiva, which date from the sixth century CE, express a vision that reaches beyond time as we know it. The god, depicted as Lord of the Dance, moves within a circle of fire that represents the continuous creation and destruction of the universe. In one hand he holds the *damaru*, the drum that he beats to mark the act of creation and the passage of time. In the other he holds the *agni*, the flame of destruction that annihilates all that the *damaru* has brought into existence. In this vision, as well as in our best scientific understanding, many phenomena are shaped by rhythm – regular patterns that cycle over time and space. 'If we speak in terms of what we perceive,' writes the linguist Vincent Barletta, 'this is merely the exposed summit of an enormous undersea mountain.'

No rhythm is more fundamental for humans than the cycle of sleep and wakefulness: the everyday miracle in which we emerge every morning into the light. And this has its origin in cycles that predate the Earth itself. When, around 4.6 billion years ago, part of a giant molecular cloud collapsed inwards, the combination of gravity, pressure and magnetic fields created a spinning disc of gas, dust and rock. Over time, about 99.8 per cent aggregated in the centre and became the Sun, while most of the rest became the wheeling planets – each a little whirlpool which rotated faster as it concentrated into a local centre, like an ice skater or ballerina who pulls in their arms.

At the time it formed, the Earth was spinning so fast that a single cycle of day and night may have taken between four and six hours. Our planet has been slowing down ever since, but it still rotates so furiously today that, at the equator, rock, ocean and atmosphere are revolving at about 460 metres per second – about 30 per cent faster than the speed of sound through air. At present, the Earth is slowing by around 1.7 milliseconds (thousandths of a second) per century, but this rate of change is imperceptible to living systems, and life finds itself on a planet with almost completely regular rhythms of light and dark.

Cyanobacteria, the first organisms to harvest sunlight by means of photosynthesis more than 3 billion years ago, evolved circadian clocks – biochemical oscillators – in order to synchronise their metabolism with solar time. They spread across the planet, releasing huge quantities of oxygen into the oceans and atmosphere, and made possible the evolution of life as we know it. These tiny organisms, which are also known as green-blue algae, continue to be foundational to life on Earth, having adapted over time to the slowly lengthening day.

Like almost everything that does not live deep in the ocean or within the rocks, humans contain molecular mechanisms like those of cyanobacteria for tracking day and night. There are also strong daily rhythms in our body temperature, blood pressure and hormones. Circadian rhythms govern the release of chemicals that regulate wakefulness, mood, the activity of immune cells, and the body's response to food. A reduction in the amplitude of these rhythms, or their disruption, is associated with poor sleep and many illnesses.

It is possible that some of the earliest time-keeping mechanisms in the bodies of our distant ancestors evolved not as circadian clocks responding to light but as circum-tidal clocks responding to the movements of the sea. The tides – ebbs and flows on the shore that are part of a long, fast, shallow wave swirling around the ocean basins in response to the pull of Moon and Sun – also create rhythms. And, driven by these rhythms, organisms such as seaweeds and the animals that live on them have evolved into complex forms and organised into semi-ordered, interlinked zones. The artist Signe Lidén and the biologist Arjen Mulder suggest that there is a sense in which these organisms are the 'thoughts' of the tide – the product of its action over billions of years. For her project The Tidal Sense, Lidén installed a twenty-eight-metre canvas – a giant version of the vibration-sensitive membrane of a microphone – on the shore of the Lofoten Islands in northern Norway and recorded the noises above and below water. She hears, among the other things, a deep, long-range protopulse resonating in all other sounds.

The acoustic ecologist Gordon Hempton suggests that we try to bring to the mind's ear the wave of song circling the Earth as daybreak sweeps from east to west and the dawn chorus of birds begins across each continent and island in turn. To this one might add two more kinds of wave. The first is the tiny, almost imperceptible sound arising from blue-green algae and other plankton as, at every sunrise, they begin to photosynthesise, and to produce little bubbles of oxygen that rise to the surface and burst in clicks and pops. The second are the many waves of sound as tides ebb and flow. Water pushes or sucks through rocky channels. Animals gather and depart. In northern waters with strong ebb tides such as those around the British Isles, dolphins, otters, gannets and other predators hunt prey that in turn are feeding on organisms flushed out by the receding water.

The tides can be locally complex, with land barriers impeding them so that they fall out of phase with the global trend. Around Britain they move clockwise, and give us, almost literally, a clock of tides, with a high tide reaching Cornwall on the south-west coast about five hours before it does Orkney off the north-east coast of Scotland, and twelve hours before Humber on the north-east coast of England. A skilled kayaker, such as the historian David Gange, can navigate at night by the sound of the tides that surge and tear around the great rocks, cliffs and sea caves off the west coast.

There are planetary rhythms over longer time frames than the succession of day and night and the tides. The tilt of the Earth's axis with respect to the path of its orbit (which may have been caused by the collision of the proto-Earth with a Mars-sized planet early in its existence) means that for half the year the northern hemisphere leans towards the Sun, which warms it, and for the other half it leans away. The seasonal change is visible in the yearly advance and retreat of snow and vegetation across the northern hemisphere. In time-lapse images from space these look like the pulses of a mighty heart or the movement of the planet's lung. 'I think of the ... land breathing,' writes Barry Lopez in *Arctic Dreams*. 'In spring a great inhalation of light and animals. The long-bated breath of summer, and an exhalation that propel[s] them all south in the fall.'

'Seasons [are] our oldest metaphors, [and] also moods, structures of feeling,' writes the poet Louise Glück, and there can be reassurance in their yearly return. Or if not reassurance, then at least a sense of wonder at what is passing such as is expressed in Gerald Finzi's setting of the poem 'Proud Songsters' by Thomas Hardy. In this music is a sense of wonder at the transformations of life, as the piano that has been lyrical and effusive disappears, leaving us in silence to absorb what has come into being and passed away.

And there are even bigger rhythms. The axial tilt of the Earth with respect to its orbit that creates the seasons itself swings between 22.1° and 24.5° over a period of about 41,000 years. Together with other cycles, such as the variation in the distance of the Earth's orbit from the Sun caused by the gravitational fields of Jupiter and Saturn, this gives rise to repeating sequences of warmer and colder regional climates over periods that range from tens to hundreds of thousands of years. Some of the consequences are visible in the repeating sequences in rock strata known as cyclothems which can make a slab of stone look like a multi-storey cake. The writer Adam Nicolson sees this 'ever-present library of time' in the rocks on the north coast of the Sound of Mull in Scotland, where layers of grey-blue limestone, lime-rich clay and black shales were laid down over and over again for millions of years in seas of the early Jurassic in 'one of the slowest of all the Earth's many songs'.

'In geological time we barely exist,' said the poet and novelist Jim Harrison. But it exists in us. These rhythms of night and day, season, tide and long-term change inform our own, and the way we perceive and live. There is, as the ecologist Aldo Leopold put it, a vast, pulsing harmony – its score inscribed on a thousand hills, its notes the lives and deaths of plants and animals, its rhythms spanning the seconds and the centuries.

The Loudest Sound

One of the loudest sounds in the Earth's recent history – if all of that history were one year, it was the day after Christmas – was caused by the asteroid which struck near Chicxulub in the Yucatán Peninsula in Mexico 66 million years ago. This was the one which (perhaps in conjunction with massive volcanic activity) killed off about three-quarters of known species, including the dinosaurs.

The asteroid, which was more than ten kilometres (six miles) across and weighed over 10 trillion tonnes, was travelling at about twenty kilometres (twelve miles) a second. It was a rock larger than Mount Everest hitting Earth twenty times faster than a bullet. In the fractions of a second before it struck the ground, the asteroid compressed the atmosphere below it so violently that it became hotter than the surface of the Sun.

The impact itself was equivalent to about 100 million megatonnes of TNT – or 2 million times the largest ever thermonuclear weapon test. Almost instantaneously, it made a hole thirty kilometres (nineteen miles) deep and 100 kilometres (sixty-two miles) wide. Over the next few seconds the Earth's crust wobbled like a pond after a rock has been thrown in. In a few moments, the peaks and ripples around the impact site rose to become a mountain range as high as the Himalayas.

Pressure waves from the blast expanded concentrically around the planet. Jay Melosh, a scientist studying the impact, offers a vision in miniature of what this might have been like. He witnessed an air blast of just a few hundred tonnes of high explosive from a kilometre away. 'You can see the shock wave in the air,' he told the journalist Peter Brannen. 'It looks like a shining [soap] bubble that expands in complete silence ... very very rapidly until you hear the kaboom! But before you hear that you feel the shaking in your feet because the seismic energy propagates faster than the sound.'

Another way to get a sense of the magnitude of the Chicxulub impact is to compare it to the loudest sound for which we have earwitness accounts. On 27 August 1883, the eruption of the volcano Krakatoa pushed a tsunami forty-five metres (almost 150 feet) tall on to the shores of Java and Sumatra some thirty kilometres (nineteen miles) away, killing between 36,000 and 120,000 people. The captain of the Norham Castle, a ship sixty-four kilometres (forty miles) from Krakatoa at the time, wrote that the explosions were so violent that the eardrums of over half his crew were broken.

More than 160 kilometres (100 miles) away, the sound of Krakatoa was measured at around 172 decibels. That is over eight times as loud as the threshold for pain in humans, and about four times as loud as a jet engine when you're standing right next to it. Krakatoa was audible 2,100 kilometres (1,300 miles) away in the Andaman and Nicobar Islands ('extraordinary sounds ... as of guns firing'); 3,200 kilometres (2,000 miles) away in New Guinea and Western Australia ('a series of loud reports, resembling those of artillery'); and 4,800 kilometres (3,000 miles) away in the Indian Ocean island of Rodrigues, near Mauritius ('like the distant roar of heavy guns'). On the other side of the world, beyond where the eruption could be heard, weather stations measured a spike in air pressure in the hours after the explosion as the sound waves spread. These pressure waves – whispers of the blast – circled the globe three to four times in each direction over a period of about five days, taking around thirty-four hours to travel around the planet each time.

The Chicxulub impact was half a million times more powerful than the eruption at Krakatoa, and the noise and mayhem it caused were correspondingly vaster. But amid the aftershocks and debris life eventually found new ways to thrive.

The Northern Lights

Old stories about the Northern Lights, or aurora borealis, show the full play of human imagination at work across the sky. In Greenland some said the lights were the spirits of children who had died at birth but were now dancing in the heavens. Others said they were made by spirits playing ball with the skull of a walrus – or by walrus spirits kicking around human skulls. To the Algonquin people of eastern Canada, the lights were the reflection of fires lit by their creator, Nanabozho, to remind them that they are in his thoughts. In Finland they are still known as 'fox fires' after a mythical fox whose tail sparks colourful flames in the sky as he runs across a snowy landscape.

A scientific account of auroras links our imaginations to the vastness of space and the irreducible strangeness of matter. They occur when charged particles of the solar wind streaming across tens of millions of miles of empty space from the Sun are drawn down along the lines of the Earth's magnetic field where it is nearly vertical at the polls. Here, 100 kilometres (sixty-two miles) or so above our heads, the particles excite air molecules which then shed their excess energy in the form of light: green or red for oxygen, blue or purple for nitrogen.

More puzzling are reports of the aurora making sounds. The explorer Knud Rasmussen wrote in the early twentieth century that the Inuit sometimes heard whistling, rustling and other sounds as it played across the sky. According to their legends, he added, if you returned the whistle the light might come nearer and even dance for you. Some Europeans reported hearing the noises too. In an account of travels in Lapland published in 1827, the naturalist Sir Arthur de Capell Brooke described one such occasion. 'The lights were ... very bright and exceptionally fast in motion ... The night was perfectly calm and quiet and I thought I heard a crackling sound ... coming from [their] direction.' Other accounts over the years compare the noise to the swish or rustle of a silken skirt, the sizzle of bacon in a hot pan, a flock of birds, and even to the crack of a rifle being fired.

For a long time many researchers dismissed these reports. Auroras occur far too high up for any sounds they might make to be audible on the ground so there could be no sound. But in 1990 a young acoustic scientist had an experience that helped change that. Taking a break during a jazz festival in the far north of Finland, Unto Laine and a friend stepped out into the cold night air. In the remote location and with no wind they expected almost complete silence, but instead heard a hissing noise above their heads that seemed to fluctuate with the movements of an aurora.

Laine half-forgot the experience but when he returned to the festival in 1999 and heard the same sounds again he decided to investigate. After dozens more observations, and after making the first recording of the elusive sounds in 2012, Laine thinks he can explain what is going on. The noise that can be heard on the ground, he says, is caused by a corona discharge – the same phenomenon that can create a blue glow around high-voltage electrical equipment such as a power line, and which is sometimes accompanied by a buzzing sound. To generate the voltages required, high positive and negative charges must accumulate close to each other. Laine reckons this can happen above our heads on unusually cold evenings when the frozen ground cools the air immediately above it. The cold air becomes trapped under a layer of warm air a few hundred metres up at the base of what is known as the inversion layer. Negative ions close to the ground rise up to this layer, but do not rise above it. Meanwhile, positive ions settle on its upper surface. This electric potential, already significant, is increased further by the aurora, says Laine, to the point where a sudden discharge happens, producing ultraviolet radiation, magnetic field pulses and sounds. All this is happening just a few tens or hundreds of metres above our heads – far below the visible aurora itself, which is typically 100 kilometres (sixty-two miles) or more up.

The sounds, which Laine describes as crackles and muffled bangs, only last for fractions of a second and are typically in the range of twenty to forty decibels, or roughly the same volume as a human whisper. Occasionally, however, they can rise to around sixty decibels – as loud as ordinary speech a few metres away. They occur most often when a geomagnetic storm is particularly strong and the air close to the ground is still and very cold, but people often miss them, Laine tells me, because they are chatting or taking photos. 'One has to listen very carefully to hear them and to distinguish them from the ambient noise,' he says.

The composer Sam Perkin has written a piece that portrays Laine's quest and discovery. In *Alta for Two String Trios and Electronics*, shimmering harmonies on the strings build slowly to a point when recordings of the auroral sounds break and crack over their surface like celestial percussion. The piece was first performed early in 2020 in the Northern Lights Cathedral in Alta, Norway – a building in which tall, slim, irregularly placed windows light up the wall behind the altar in a way that resembles the aurora. The Northern Lights – 'the drum taps of a solar storm upon the Earth' – were echoed in both architecture and music.

Others, meanwhile, explore the sound worlds that can be

created when the soundless electromagnetic waves of the aurora itself are transformed through a synthesiser into analogues that we can hear. The results resemble whistling whales, calling frogs and the twittering of strange birds. 'When they're very prominent, you'll hear this sweeping frequency from high to low or low to high,' says the sound artist Matthew Burtner, who has used the sounds to create a piece titled *Auroras*. 'You really feel like you're in touch with the solar system.'

Volcano

A monster bigger than you ever thought it possible for a monster to be is humming to itself in a vast cavern. A gamelan at the bottom of the ocean is joined by extraterrestrial beings playing electronic piccolos. Such are the sounds of a dormant volcano as detected by the Infrasound Laboratory at the University of Hawai'i. The lab's primary mission is to listen out for human activity that would break the Comprehensive Nuclear Test-Ban Treaty, but it also compiles noises made by the planet itself, including these subterranean rumblings of the volcano Kīlauea on the big island of Hawai'i. Unprocessed, the sounds are so deep that humans cannot hear them. But sped up one or two hundred times they can be stored as audio files in an ever-growing library of the sounds made by our planet.

Sounds and vibrations from tectonic and volcanic activity are among the most powerful and pervasive sounds on, and within, the Earth. A visualisation created by the Seismic Sound Lab at the Lamont Doherty Earth Observatory called the Seismodome shows how most earthquakes ripple out in concentric circles, setting the Earth ringing like a bell. Hydrophones near Ascension Island in the South Atlantic can detect submarine volcanoes on the other side of the planet. Volcanic eruptions can make some of the loudest known sounds on Earth that humans are likely to hear, but their most intense vibrations are typically in the infrasound range, far below our hearing. These wavelengths, which are typically around one hertz (that is, one wave per second), and hundreds of metres in length, are largely determined by the dimensions of volcanic craters, which modulate sound just as the size and shape of the horn of a musical instrument such as a trumpet help determine its pitch and timbre.

For decades volcanologists have monitored infrasound to count explosions and to track eruption intensity. More recently they have also begun to listen to volcanoes before they erupt, monitoring the changing character of the noise with the aim of better predicting what will happen next. Typically, there may be around fifty volcanoes around the world in 'continuing eruption status' at any given time and of these about twenty, anywhere from Antarctica to Iceland and from Japan to Peru, are likely to be actively erupting on a given day. They make lots of noise, but they also create other kinds of waves besides sound waves. The Tonga eruption of January 2022 triggered ripples in solid rock that travelled several times around the Earth at hundreds of kilometres per second, and gravity waves that extended to at least the edge of space.

Volcanic eruptions make a variety of sounds which fall within the range of our natural hearing too, and for thousands of years people have found ways to talk about them and the noises they make. In what may be the world's oldest story, the Gunditjmara people of South-east Australia tell of a giant whose body transformed into a volcano called Budj Bim, which spat out his teeth in the form of lava – an allusion to actual events their ancestors observed 37,000 years ago. Witnesses to the dramatic explosion of Volcan Fuego in Guatemala in the early eighteenth century talked variously of *ronroneo* (purring), *retumbo* (rumbles), *bramido* (roars) and *estruendo* (booms/blasts). The volcanologist David Pyle told me that some of the strangest sounds he has heard in decades of fieldwork were those made by Ol Doinyo Lengai, a 'bizarre and remote' volcano in Tanzania which one misty morning in 1988 emitted 'chuffs, puffs, bursts and hoots' that sounded like nothing so much as a Victorian railway station. Similarly gentle sounds were heard at the Fagradalsfjall volcano in Iceland during its eruption from spring to autumn of 2021. This was a bijou eruption – effusive rather than explosive – and for much of the time one could walk safely to within a few tens of metres of the slowly emerging lava. The volcano became a local and international tourist attraction, with people having picnics and even get married next to it. The writer and film-maker Andri Snaer Magnason described the soundscape made by small streams of lava as soothing and seducing, as if the volcano were whispering 'come closer'. In a clip that he posted on Twitter, heather blooms untouched just a few metres from bright orange-red lava. 'Kali, doing her thing,' he wrote, alluding to the goddess who is associated with destruction but who also protects and preserves the natural world.

When I spoke to Magnason about his experience at Fagradalsfjall he emphasised how soft and gentle the sound had been – something like a woofer, or a kind of deep breathing. 'It was almost like a pet volcano from *The Little Prince*,' he said. He also wrote that he had felt a connection to forces beyond normal experience, a sense of 'the planet, the origin of life, the creation of everything, the recycling of materials and continents, the elements that make the planet what it is . . . In the end, the scene [has] nothing to do with terror . . . or hell. It [is] destruction and creation happening simultaneously.'

The writer Heidi Julavits came to a similar conclusion on her visit to Fagradalsfjall that spring – although in her account, the sound of lava varied from a thick sloshing noise like a stomach loudly digesting to an oceanic roar. 'When I'd sent my husband videos of the smoking lava fields,' she wrote, 'he'd texted back, "It's like watching a city get bombed."' Julavits replied that for her 'the scene evoked the opposite of annihilation'. But upon reflection, she concluded that both responses are accurate: 'a volcanic eruption collapses the distinction between ruin and progress'.

Fagradalsfjall was like a pinhole view of a vastly older and larger picture. Without volcanoes, most of Earth's water would have been trapped in the crust and mantle billions of years ago, no land would have risen from the oceans, and there would be no breathable atmosphere. So when you listen to a volcano you hear sounds of processes that enabled life on Earth to exist in the first place. You also hear the planet's most destructive endogenous force, as when 251 million years ago, at the end of the Permian period, volcanic activity increased sharply and carbon dioxide emissions from the eruptions rapidly raised global temperatures, resulting in the extinction of some 70 per cent of species on land and 80 per cent in the oceans. (Human activity could exceed this in destructive impact if our current rate of greenhouse gas emissions, which is about ten times that of volcanoes in the Permian, continues for any length of time.)

The sound artist Jez riley French argues that 'durational listening' – sitting still and paying attention for a long time – 'can allow [a] place to impose itself on us, outside of the limits of our normal attention'. I didn't get a chance to visit Fagradalsfjall when it was erupting, but the idea of sitting beside a volcano (in reasonable safety) appeals for exactly this reason. I would like to hope that listening to the rumbles and breaths and sloshing sounds at the open crust of the Earth could be a way of extending my attention, just as lava itself stretches and extends into new forms.

Thunder

'Hear, O hear His voice raging,' says the Book of Job of the god Yahweh. 'He thunders in the voice of His grandeur.' And what is said of Yahweh is also said in various ways and times of gods and spirits across six continents and the islands beyond, who announce both destruction and life-giving rains with rumbles through the sky. Some of their names are Teshub, Hadad, Set, Tarḫunna, Zeus, Taranis, Horagalles, Orko, Perëndi, Perun, Thor, Leigong, Raijin, Takemikazuchi, Indra, Wakinyan, Kw-Uhnx-Wa, Hé-no, Yopaat, Xolotl, Chibchacum, Tupã, Shango, Amadioha, Xevioso, Sudika-mbambi, Kanehekili, Mamaragan, Ipilja-ipilja, Whaitiri and Tāwhirimātea.

The ubiquity in human imagination of storm-beings who speak in thunder should be no surprise. Thunderstorms strike somewhere on Earth between forty-five and a hundred times a second, or about 3 to 8 million times a day. They occur everywhere except the polar regions, which are usually too cold for sufficient convective heat to accumulate in the atmosphere and set them off – though this may be changing as the planet heats up: in the summer of 2021 meteorologists were astonished to see three thunderstorms sweep across from Siberia to the north of Alaska in a single week.

In a thunderstorm, hail or ice falling from the upper part of the storm collides with water droplets rising from lower down, and its surface is slightly heated so that it becomes soft hail known as graupel. As this graupel collides with more rising droplets it shears electrons off them and carries the electrons (which carry a negative electrical charge) downwards. Over time, negative electrical charge accumulates in the lower part of the storm or at ground level while positive charge accumulates in the upper part. Air is an excellent electrical insulator, so the charge separation between the two parts has to be extremely high (in the millions of volts) before lightning will discharge the imbalance.

Lightning is, as a wag at the UK Met Office puts it, 'the width of your thumb and hotter than the Sun'. The narrow channel of air through which it travels heats up almost instantly to as high as 30,000 °C – about five times the temperature of the surface of the Sun. This causes the air to expand extremely rapidly and creates an acoustic shockwave: thunder is the sound of exploding air. Even at its loudest, though, the sound accounts for only about 1 per cent of the energy from the discharge while 9 per cent is light and 90 per cent is heat.

Nearby, thunder makes a sharp crack or loud bang. More distant thunder rumbles and ripples longer and lower because the listener is hearing more from along the line of the lightning and because higher frequencies in the original noise are absorbed more quickly by the intervening air. Thunder spans a wide range of frequencies. The crack is usually in the range of 2,000 hertz and followed by lower harmonics from about 600 to 800 hertz and a chest-rumbling resonance mostly from seventy to eighty hertz, with a low end at forty to fifty hertz. Reverberation and temperature differences in layers of the atmosphere can add an echo or delay that makes the sound seem bigger and deeper.

Thunder can be about 120 decibels when you're really close. This is about as loud as a pneumatic drill or standing right in front of the biggest speakers at a rock concert in full blast. I generally try to avoid this kind of thing, but I do know what thunder sounds like when it is almost on top of you. Lost one night high on a mountain in Nepal, I tried to hide from a thunderstorm in a crevice under a big rock (which, I learned afterwards, is just about the most dangerous place to be, and the stupidest thing to do). Facing into the rock with my eyes closed and my head covered, everything became blinding light for an instant and every part of me shook with a huge crash. 'My heart trembles,' says a verset in Job in reference to thunder, 'and it leaps from its place.' I can still feel that crash more than twenty-five years later.

'God thunders wondrously with His voice,' says the Book of Job, 'doing great things we cannot know.' But post-Bronze Age science has revealed some of them. Plants need nitrogen from the atmosphere to grow and, while much of it is fixed for them by bacteria and fungi, they also benefit from thunderstorms. The extreme heat of lightning joins the normally unreactive nitrogen to oxygen to make nitrates, and these mix with water to fall as nourishing rain.

Thunder may also have been the voice of processes that, along with volcanic activity, helped make life possible in the first place. All life on Earth depends on phosphorus, an element that is a crucial part of the double-helix backbone of DNA and the related molecule RNA. Phosphorus was widespread on the young Earth around 4 billion years ago, but it was mostly locked up in unreactive and insoluble minerals. While meteorites brought some phosphorus in reactive forms to the planet, it looks as if lightning strikes also played a key role by freeing significant quantities of the element from the rocks already present on Earth.

The thunder and lightning that accompany many volcanic eruptions today when their plumes create an electrical imbalance with the surrounding atmosphere would also have been

Listening to a Rainbow

In a poem, Douglas Dunn describes how he chances upon notes and sketches he had made as a young man and finds among them a first line that had never gone any further: 'It is like listening to a rainbow.' The simile is striking – and is simultaneously beautiful and nonsense.

The beauty in the line is clear to see. One does not need the synaesthesia claimed by the painter Wassily Kandinsky, who said that for him music and sensations of colour were inextricable, to have had an experience in which strong emotions aroused by a sight or a sound give rise to an internal sensation of the other. This can even happen to someone as grumpy and uncooperative as me, as it did when I went to a gong 'bath', in which you sit or lie in front of a gong and allow the sound to wash over you. Despite my prejudices and a mind that jumps around like a monkey on amphetamines, the vibrations rippling gently through my body created a sense of well-being, and I began to see in my mind's eye a great sheet of water flowing down over a rock face in bright sunshine.

But there is also a way in which Dunn's line makes no sense at all. Isaac Newton and other natural philosophers tried to map the colours of light on to sound but failed because the two are completely different physical phenomena. Sound is a wave of pressure passing through matter – a jostling and jiggling of atoms as they pass on vibration to their neighbours. Light, by contrast, is an electromagnetic wave, and as such it can pass through a vacuum, which sound never can do. The light in a rainbow makes no sound.

Even so, analogies and metaphors can enrich and deepen both poetic and scientific appreciation without blinding us to any truth. So, for example, some behaviours of light can be explained when it is described as a 'wave'. Two waves of light can be configured so that they interfere with each other to create darkness. Similarly, two waves of sound can be configured to create silence.

One can also talk about 'colours' of sound on the basis that light with the same power spectrum (that is, the same relationship between the frequency and intensity of its waves) would appear to be that colour. The best-known example is white noise: the 'static' on a radio or old-style TV. This is analogous to white light in that it contains all wavelengths at equal intensity. An acoustic engineer will happily bend your ear about a whole spectrum of other colours of noise including blue, violet, red and green. Most of these are human creations, present only in an electronic sound system or a computer. But there is at least one major exception, and it brings us back to the rainbow.

In what is known as pink noise, the intensity and frequency of the sound are inversely proportional so that the sound gets progressively quieter as it gets higher. Pink noise follows the same profile of statistical fluctuation as many natural phenomena, from quasar light emissions to the changing heights of tides and rivers, and from heartbeats to the firing patterns of neurons. And like many of them, it is fractal, which means it is self-similar at different scales: the pattern repeats whether you zoom in or out. Pink noise also shares this property with human music to some extent (although music almost always includes variations within its overall structures, harmonies and rhythms). And like music (but unlike the harsh electrostatic hiss of the white variety), it is pleasant to human ears. It is very like the sound of a waterfall, a breaking wave or a fine rain. Some studies claim to find a positive link between exposure to pink noise (and its close cousin, brown noise) and restful sleep.

'Now it can be heard,' says Douglas Dunn in another poem titled 'Wondrous Strange'. But almost immediately he corrects himself: whatever sound seemed to be there is 'not quite' but is 'still on the far side of nearly'. And so I like to think of the sound of a rainbow. If you see a rainbow from afar and, quickly, while it is still raining, go to stand in the place it appeared to be, you will of course not see it, because the reflection and refraction of light in water drops that create a rainbow require distance. But you will be in the midst of one part of what – together with the light and the human witness – co-creates the rainbow: a myriad fine drops of water, gently falling. And maybe, if you listen very carefully, you may hear those drops, and know that for someone elsewhere you too are part of the rainbow.

BIOPHONY

Sounds of Life

.

Rhythm (2) – Body

Heart: 'rhythm is, first of all, the rhythm of the organism, ruled by the heartbeat and circulation of blood,' wrote the poet Czesław Miłosz. He was not entirely right: our bodies follow circadian rhythms – the planet waves of day and night – that predate the evolution of the first animal heart around 520 million years ago, but the heart's steady beat is the click track of our lives and it pulses more than 3 billion times in a life of eighty years. At rest, the rate can fall below fifty per minute. During exertion or fever, it can exceed 200. The range corresponds quite closely to the range of tempos in music and dance. Researchers have found that people who are more aware of their own heartbeat are better at perceiving the emotions of those around them.

Breath: for Rainer Maria Rilke, breath was an 'invisible poem / pure exchange . . . rhythm wherein I become'. The journalist James Nestor reports that rhythmical breathing practices associated with acts of prayer and meditation in many cultures can lead to significantly better health; ideally the in- and out-breaths last for five to six seconds each. But rhythmical breathing also plays a central role in some of the most ordinary activities. When people engage in friendly conversation, says the neuroscientist Sophie Scott, they immediately start to breathe in synchrony, and match each other's rhythm and pitch.

Footfall: kangaroos bounce and blue-footed boobies strut, but humans are the only animals that walk and run upright on two legs over long distances. Both movements are a kind of controlled fall, in which each leg swings forwards in turn to stop us hitting the ground and then powers us along. (Over and over you're falling, but then catching yourself from falling, sings Laurie Anderson.) Walking and running on two feet require an exquisite sense of balance and, the psychologist Aniruddh Patel suggests, learning to do so may have helped develop our ability to measure and predict more than just bodily movements and to refine our awareness of time and rhythm. Dance, song and a feel for the length of a journey may start with the regular beat of our steps.

Brain: the brain has several ways of counting time, from the interval timer, a network of neurons that registers durations from seconds to hours, to the suprachiasmatic nucleus, a small region of the brain that synchronises body processes to the passage of day and night. These, and many other functions, are supported by brainwaves – repetitive patterns of oscillation, or the firing of action potentials in neurons – across many cells together that range from about 0.02 to 600 cycles per second. 'Brains are foretelling devices and their predictive powers emerge from the various rhythms they perpetually generate,' says the neuroscientist György Buzsáki. Because of a non-integer relationship among their frequencies, he says, the oscillations can never perfectly entrain – lock into step with – each other. Instead, the interference they produce 'gives rise to metastability, a perpetual fluctuation between unstable and transiently stable states, like waves in the ocean'. Buzsáki compares the multiple timescale organisation of brain rhythms to Indian classical music, where 'multilevel nested rhythmic structure, explained by the concept of tāla, characterises the composition'.

Hearing

I sometimes find small snails in our vegetable patch. Usually I remove these characters, but sometimes I examine one more closely before I do and marvel at the minuscule retractable stalks tipped with eyes, the perfectly formed shell, and the near weightlessness of the whole. When they are the size of a pea, the shells of these snails are about as big and roughly the same shape as the cochlea: that part of the bony labyrinth of our inner ears where vibrations in the air around us are turned into what we experience as sound. Through two of these structures, the booming and buzzing confusion of the world, all its voices and music, passes into the kilo and half (or three pounds) of wobbly blancmange inside the nutshell numbskulls that are our kingdoms of infinite space.

The more you examine it, the more hearing seems like a superpower. In her memoir of how she lost and then regained it, Bella Bathurst remarks on the subtlety and discrimination with which those with normal hearing can attend to a human voice: 'the exact way your husband says "brilliant" or "orchestra" ... The way your son's first cries differed from your daughter's ... What your boss sounds like when she's nervous ... Your partner's voice down to ... several thousand shades of "no"'. This capability enables humans to cooperate, empathise with and manipulate each other in ways that few other species match – though whales, parrots and some others should not be underestimated. It also enables some of us to 'read' aspects of the non-human world with amazing precision. Spend time in a forest with someone who attends to birdsong and they may reveal diversity and detail that you might scarcely have guessed at before. The ability can be no less remarkable in a technological context. The journalist George Monbiot describes a friend who, just by listening over a phone line, can diagnose faults in a car engine with a precision that astonishes professional mechanics.

Touch is often regarded as the most primal sense. It 'orients us to the fundamental condition of being – to the inevitability of others, both human and non-human,' observes the author Nikita Arora. But hearing begins before we are able to touch and be touched by the world around us, and so before the distinction between self and other has formed. The inner ear develops rapidly in the foetus, reaching adult size by the fifth month of gestation, and passing information to the temporal lobe of the brain, which processes sound. Low-frequency sounds travel best into the womb, and babies start to form memories of sounds while still inside their mother's body.

Hearing is also often our most rapid and agile sense in adult life. Sprinters react faster to a starting pistol than to a visual cue such as a flag even though the light travels nearly 900,000 times faster than sound and reaches their eyes before sound reaches their ears. Hearing is also often the last sense to go at the end of life. The dying brain can register sound, even in an unconscious state, up to the last hours.

Hearing originated in the evolution of tiny hair-like structures on the outsides of single-celled organisms hundreds of millions of years ago. The cells moved some of these 'cilia' (from the Latin for eyebrow) from side to side to propel themselves through water, and used others to sense vibration and movement in the water, or contact with another object. They also passed both types – motile and non-motile – down to the descendants. The former are still in us today on cells that line our lungs and respiratory tract, which push mucus and dirt up and out of the body. And the latter found a place on the lateral lines of our ancient fishy ancestors and on most fish today. These lines are rows of tiny cups along the outside of a fish's body which house thousands of cilia and are exquisitely sensitive to small movements and vibrations. 'It is not too much,' writes the philosopher of science Peter Godfrey-Smith, 'to say that a fish's body is a giant pressure-sensitive ear.'

But one kind of ear is not enough for many self-respecting fish in a dynamic and often dangerous environment. Their and our aquatic ancestors also evolved inner ears, in which the cilia sense movement not in the water pushing against their bodies but in otoliths: tiny lumps of calcium carbonate inside their heads. Fish are about the same density as water, but the otoliths are denser and so move at a different amplitude and phase in response to vibrations passing through the fish's body. The cilia register this, and the fish interpret it as sound. Some fish have also evolved additional ways to boost their hearing by using their swim bladders – gas-filled sacs whose main role is for buoyancy – as amplifiers or hearing aids. Thanks to these swim bladders, species such as the American shad can hear sounds up to 180,000 hertz, nine times higher than humans, and detect the ultrasonic vocalisations of the dolphins that want to eat them. Others can hear infrasound far below the lowest threshold for humans. This allows them to sense the rumble of the tides and the movement of water as it breaks and flows around rocky shores, and helps them in navigation. In the deepest waters, 8,000 metres down, snailfish detect amphipods through subtle vibrations in fluid-filled chambers in their jaws.

Hearing on land presents new challenges. The lateral lines filled with cilia that run over a fish's body don't work in the much thinner medium of air, and sound waves passing through the air are muffled when they enter the much denser body of an animal. Even so, some of the early tetrapods – the first vertebrates to have limbs and digits – may have been able to hear above water without the complex apparatus that evolved in their descendants. Modern lungfish, which resemble those first venturers on land, hear by using their lungs as amplifiers rather as some fish use their swim bladders underwater. It's a start, though a fairly basic one, and fully terrestrial creatures have had more than 380 million years to evolve improvements.

Part of the story begins with the spiracles: little canals on the top of the heads of air-breathing fish through which they sucked in oxygen from the air above the murky water that they slithered through. Early amphibians adapted these spiracles into tympanums – tubes with membranes stretched over the top such as you can still see in the circles just behind the eyes of frogs. In frogs, the tympanum – a simple eardrum – is joined by a tiny bone called the columella to the inner ear where the sensitive cilia are located. It is also connected by an open passage to the lungs so that a frog can equalise pressure on both sides and not deafen itself with calls to its friends, which in the case of the common coquí of Puerto Rico can exceed 100 decibels – about as loud as a power mower right next to you, or a jet taking off 300 metres away. In mammals, the spiracles have evolved into the Eustachian tubes that connect the middle ear to the nose and throat. Next time you equalise pressure in your ears by blowing out while closing your mouth and pinching your nose shut, thank Tiktaalik, your inner fishapod.

Another part of the story begins with a transformation as strange as anything in the *Metamorphoses* of Ovid: an act of incredible shrinking origami as, over time, parts of the jaws of our ancestors evolved into a tiny, elaborate gearing mechanism. Starting out as a long bone that braced the lower jaws of early fishes, the hyomandibular became what we now call the stapes, while the quadrate and articular bones, which brace the lower and upper jaws in fish, became the malleus and incus. The Latin names of these three 'ossicles' – little bones which in humans are about the size of mini-Lego – comes from their resemblance to a stirrup, hammer and anvil. Together they are articulated like a set of levers. One end of the malleus connects to the centre of the inside of the eardrum. Vibrations striking the eardrum from the outside cause it to move, and its inner end moves the incus, which in turn prods the stapes. The inner end, or faceplate, of the stapes rests against the cochlea through what is called the oval window. The eardrum has an area of about fifty-five square millimetres while the faceplate of the stapes is about 3.2 square millimetres, so the vibrations in air are concentrated into a larger movement over a smaller area. This enables the relatively weak vibrations in the air to pass into the denser, fluid-filled interior of the cochlea. If a jaw is – in a happy phrase of Peter Godfrey-Smith – a thumb for your face, then the bones from our jaw that evolved to be the ossicles are a car jack for your ear. The mechanism is so sensitive in humans that we can sense vibrations such as those created by an ordinary speaking voice that momentarily change atmospheric pressure by just a few billionths. According to some calculations, the quietest audible sounds move the eardrum by a distance of less than one picometer – that is, a thousandth of a nanometer, or about a sixtieth the diameter of a hydrogen atom.

It is not essential to have architecture as elaborate as this, however, to have excellent hearing. Birds, whose outer ear opens to the air as a flat hole that is usually concealed by feathers, resemble amphibians and reptiles in having only one bone in the middle ear, and their cochleas are relatively simple straight tubes. Despite this, many birds hear at least as well as or better than humans over roughly the same range of pitches – albeit with a peak in sensitivity in a higher part of that range from about 1,000 to 4,000 hertz. And some can detect far lower sounds. Pigeons may be sensitive to infrasound as low as 0.05 hertz, perhaps enabling them to navigate long distances by attending to vibrations generated by deep ocean waves and earthquakes. Humans typically hear no lower than about twenty hertz unless a sound is very loud indeed. Further, all birds have an ability that humans lack: they can regrow the cilia in their inner ears, while we lose them for good when they are damaged. And the brains of many songbirds can process sounds as much as ten times faster than we can, enabling them to follow complex sequences of different tones where we hear only a blur. If only they could understand English they'd make good students for the geneticist Steve Jones, who used to joke that the problem was not that he spoke too quickly, but that his students listened too slowly.

Almost all mammals, from the smallest (the two-gram Kitti's hog-nosed bat) to the largest (the blue whale, which can weigh 170 tonnes), have snail-shell-shaped cochleas. The exception are monotremes – egg-laying mammals such as the platypus, which Ogden Nash felicitously but unreliably described as partly birdly, partly mammaly, and which have cochleas shaped like bananas. (By the way, a baby platypus is called a platypup. You're welcome.) In humans, the cochlea twists three times around a central axis for a total of about thirty-two millimetres, or an inch and a quarter. Unwhorled, it would be about as long as the final joint of your thumb.

Threaded along the inside of this whorl is the basilar membrane, so named because it is attached at its base to the cochlea's open end. When the vibrations reach it, this membrane starts to undulate in time. Resting on top of it are specialised neurons with hair-like protuberances sticking up, like hair on a crew cut. Lying above them is another more rigid membrane. As the basilar membrane and hair cells bob up and down in response to a vibration, the hairs scrape this 'roof' in a motion that the neuroscientist Jennifer Groh describes as being 'like kayakers banging their heads on the ceiling of a sea cave, but with more useful results'. Under the stress of brushing back and forth, pores in the cells are alternately stretched open or pushed shut. The sea of fluid surrounding the hair cells contains ions – molecules with an electrical charge. As the hairs bend in one direction, the ions flow through the tiny molecular gate, and as the hairs deflect in the opposite direction, the pores close and the motion of the ions stops, creating electrical signals that track the sound. In this way, movements of air molecules outside the body are converted into electrical signals that are transmitted by neurons in the auditory nerve to the thalamus and on to the auditory cortex in the temporal lobes of the brain, which sit inside the skull right next to the ears.

The human ear detects a range of frequencies thanks to different physical properties along the length of the basilar membrane. At its base just by the oval window the membrane is narrow – about a fifth of a millimetre across – and comparatively stiff, but as it threads into the cochlea it gets wider, like a kipper tie from the 1970s, and about a hundred times more floppy. The stiffer part vibrates in response to higher frequency sounds – for humans up to 20,000 hertz; the floppier end responds to low frequencies. It's as if we had coiled-up piano keyboards in our inner ears – or a tonotopic map, where sound becomes place. When we hear many frequencies at once, as is usually the case, several waves form along the membrane. As we age we tend to lose our hearing at the upper end of the normal range (the song of a cricket, say, or the highest birdsong) because this is where vibrations are least attenuated and because there are simply fewer hair cells to lose on the narrower strip.

Loud sounds vibrate the membrane with ease, stimulating the inner hair cells and so creating nerve impulses. In a healthy ear the cochlea will actually amplify quieter sounds by as much as a thousand times thanks to a second, more numerous set of hair cells on the other side of the basilar membrane. When a faint sound moves these outer hair cells, a protein called prestin, which is the fastest known force-generator in living cells, responds to amplify the wave, which in turn triggers the waiting inner hair cells. This arrangement allows the human ear to perceive sound across a million-fold difference in energy levels, from the fall of a snowflake to a thunderclap.

The outer ears of humans – those flaps of folded skin and cartilage on the sides of the head which we often simply call 'ears' but which are technically known as pinnae or auricles – may look odd and frankly underwhelming when compared to the huge, movable apparatus of a fennec fox, but they should not be underestimated. Our pinnae amplify sound by fifteen to twenty decibels – an effective doubling of volume equivalent to walking across a large room to stand next to someone who is talking. And they do more than this. Most sounds in the world around us include a wide range of frequencies, and some of these are dampened as they bounce off the swirling peaks and valleys of the pinnae while other waves reinforce each other and are made stronger. The folds on the pinna 'act like an equaliser on a stereo system,' says Groh, 'boosting the base or treble depending on where the sound is coming from' and preferentially select sounds in the frequency range of human speech. They also help you work out whether a sound is coming from in front or behind, above or below. The little flap of skin just in front of the ear canal, for instance, which is called the tragus, reflects and filters sounds coming from behind.

Because we have a head between our two ears (I speak for myself, at any rate), small differences in when each of them registers a sound also help us work out where it is coming from. Sound travels slowly enough through air – about 344 metres per second, or 767 miles per hour – for there to be a detectable difference in the time it takes to reach each ear. A sound directly to your left or right takes about half a millisecond – that is, half of a thousandth of a second – to reach the ear on the other side of your head and will be slightly quieter, and the brain can tell the difference quite easily. But auditory processing in the brain can do better than this, detecting time differences of as little as ten to thirty microseconds (millionths of a second). This is hundreds of times quicker, and only enough for a sound to travel from about three to nine millimetres. Thanks to this capacity, we can distinguish between two sound sources three to seven centimetres apart from each other and two metres away from us. Have someone sit across a table from you, close your eyes, and see if you can tell whether they are snapping the fingers of their left or right hand as they bring those hands closer and closer.

Capable as we are of sensing sound waves at up to 20,000 cycles per second – or about a thousand times as fast as we can resolve visual information – hearing is a human superpower, but in some other animals it is superer. Dogs can hear up to 40,000 hertz – twice as high as humans – and cats up to 80,000. Mice and rats chitter and warble to each other at up to 90,000 hertz. Porpoises hear up to 140,000 hertz; bats up to 200,000. Compared to all these animals we are sonic flatlanders, oblivious to huge parts of the sonic spectrum from the deep noise of ocean storms, earth tremors and volcanoes to what Jacquetta Hawkes called 'the fine tissue of imperceptible sounds: vegetation growing, leaves and flowers moving, all the stirrings of growth and decay'. To this one may add high-pitched insect songs, the cries of bats, and the soft fizzes and pops of water and sap travelling through the veins of plants. 'There's a poignancy in these limitations,' writes David George Haskell. 'The world is speaking, but our bodies are unable to hear much of what surrounds us.'

But if we recognise our limits – that, as George Eliot wrote, 'the quickest of us walk about well wadded with stupidity' – we may also appreciate the little that we do have, and begin to reflect on how we may understand more. 'When we have learned how to listen to trees,' writes Hermann Hesse, 'then the brevity and the quickness and the childlike hastiness of our thoughts achieve an incomparable joy.'

Ancient Animal Noises

On 29 February 1832, two months into what was to be a five-year voyage around the world, Charles Darwin stepped ashore and into the coastal rainforest of Brazil. 'A most paradoxical mixture of sound and silence pervades the shady parts of the wood,' he wrote. 'The noise from the insects is so loud, that it may be heard even in a vessel anchored several hundred yards from the shore; yet within the recesses of the forest a universal silence appears to reign.'

If this effect filled Darwin with wonder, what would he have made of the forests of the Carboniferous that spread over swathes of the supercontinent Pangaea from about 360 to 300 million years ago? Here, in great swamps, trees resembling giant ferns and enormous horsetails grew high into the sky. Amphibians the size of crocodiles and scorpions as big as Labradors picked their way through fallen logs that fungi and bacteria had not yet learned to break down. Dragonflies the size of seagulls flew overhead. What chirps, clicks and grunts – and what heavy silences – would have filled the air?

Hundreds of millions of years of life are locked away forever in the abysm of time, but scraps of evidence, combined with deductions based on the sounds that some animals make today, mean that we can reimagine and recreate some of the noises made by prehistoric animals.

Acoustic communication on land is thought to have evolved just once in a common ancestor of land animals and lungfish around 407 million years ago. The earliest note that is so far known with precision was sung around 165 million years ago by a katydid, a kind of cricket, which rubbed a serrated vein on one wing against a plectrum on the other in what is known as stridulation. The exact pitch – at 6,400 hertz, a little above a G8, or about a fifth above the highest note on the piano – can be deduced from the anatomy of remains preserved in amber. In a recreation published in 2012 it sounds tinny and electronic: a very small fire alarm running out of batteries.

Dinosaurs almost certainly used sound to communicate with each other and to listen for predators or prey. The fossilised remains of the inner ears of *Thecodontosaurus antiquus*, a nimble two-legged omnivore the size of a ten-year-old child that lived around 205 million years ago, show that it would have been able to recognise the squeaks and honks of various other animals. And *Mononykus*, a small, feathered dinosaur that lived in the deserts of Mongolia about 70 million years ago, had a facial disc of feathers very like those of a modern barn owl which would have amplified sound and funnelled it to the animal's ears just as those of the owls do today.

Direct evidence for dinosaur vocalisations, however, is scarce. One of the few cases that palaeontologists describe with confidence is *Parasaurolophus*, a plant-eating 'duckbill' dinosaur, or hadrosaur, that lived in what is now western North America around 75 million years ago. Up to four metres tall and more than nine metres long – roughly the size of a bus – this creature probably made noise through a hollow bony tube on its head that was connected to its nostrils, curled back over the top of the skull, and extended behind as far again – like a slightly flaccid upside-down didgeridoo. In the 1990s, scientists at Sandia

Labs in New Mexico (where the normal business is making parts for nuclear weapons, but where there was presumably some downtime) created a full-size replica of this appendage, and discovered that it made a splendid noise. Pitched at around thirty hertz, or just below the bottom note of a piano, the timbre has been compared to a trombone, and because of this *Parasaurolophus* is sometimes called the trombone dinosaur. To my ears, there's a dash of French horn or sousaphone and a hint of creaking metal door in there too. Perhaps, like the members of the Hinkle-Horn Honking Club in *Dr Seuss's Sleep Book*, these creatures would honk themselves out each day, but wake up fresh the next morning to start right on honking again.

Big carnivorous dinosaurs such as *Tyrannosaurus rex* did not roar like they do in the *Jurassic Park* films, where the sound is created by mixing the slowed-down bellow of a baby elephant, the roar of a tiger, some whale song, an alligator hiss and the engineer's pet dog (though not, apparently, the kitchen sink). Rather, palaeontologists conclude from study of some of *T-rex's* closest living relatives such as crocodiles and the modern avian dinosaurs (known to most of us as birds), they probably boomed, closed-mouthed like a croc, at the lower end of human hearing and into the infrasonic range, with perhaps some groaning hiss thrown in. You would have felt the vibrations through your soon-to-be-chomped bones before you heard or saw the beast that made them.

A few million years before the Chicxulub meteorite wiped out the non-avian dinosaurs, the ancestors of today's birds evolved a new kind of vocal organ. The oldest fossil of one found so far belonged to *Vegavis iaai*, a bird that lived in what is now Antarctica about 67 million years ago. Located in its windpipe just above the lungs, the syrinx enabled *Vegavis* to quack or honk, rather like its living cousins, ducks, geese and swans. Some biologists think that the evolution of the syrinx, which is the avian equivalent of the larynx, or 'voice box' in humans and other mammals, may have helped birds to survive and diversify by enabling them to communicate better with each other and engage in more complex behaviours.

At first the new vocal organ in birds was probably fairly limited in its capabilities, but over time it made possible something auraculous, or ear-marvellous. The decisive factor, it appears, was a period around 30 million years ago when the climate in Australia became dryer, and plants adapted by excreting more of the sugars they produced instead of turning them into leaves, seeds and woody material. This in turn gave rise to a phenomenon that is still visible today in parts of the continent that are not utterly parched, or incinerated by humans: Australia, writes the biologist Tim Low, has 'forests that exude energy'. Many flowers are astonishingly copious in nectar, while eucalyptus trees ooze a sweet 'manna' from their bark. Insects suck more plant sap than they can digest and excrete much of it out again as 'lerp', or honeydew. Australian birds quickly evolved the ability to taste sweetness and tap into the newly abundant supplies. And on the back of this sugar rush they transformed. The unusually large, aggressive, clever and vocal birds of the island continent today are one result, but so are all the songbirds across the rest of the world, whose ancestors spread out from these Australian roots and account for about half of the ten thousand or so species of bird alive today. As the science journalist Ed Yong puts it, 'all of these [song] birds descended from an ancestor whose voice lilted through Australian trees and whose taste buds were tickled by sweet Australian nectar'. The complex, fluting songs of nightingales, blackbirds, starlings, pied butcherbirds, robins, cardinals, thrushes, finches and many others alive today are, almost literally, a gift of ancient Australian sunlight – although one may add that in the case of birds in the honeyeater family who eat lerp, music is the poop of bugs.

The Great Rift Valley in East Africa, where humans evolved from around 2 million years ago, would have been filled with the noises, calls and songs of insects, frogs, songbirds and other animals. Cicadas, suggests the ecologist Peter Warshall, may have been humanity's first tuning fork: a reliable and constant pitch against which to gauge some of their first songs. The harmonic intervals and complex vocal signalling of birds would also have influenced our first music and language. We came to know ourselves, in as far as we do, in the context of these sounds – 'tumbling' as W. S. Merwin writes, 'upward note by note out of the night . . . song unquestioning and unbounded'.

Plant

'It is not the trees . . . that make a wood,' the author J. A. Baker once suggested, 'but the shape and disposition of the remaining light, of the sky that descends between the trees.' Something similar may be true in relation to sound, because the forest also becomes apparent to many sentient creatures through the resonances in the spaces between the trees. The tap of raindrops on leaves, the clack of branches and the rustle of vegetation shape our sense of what surrounds us. According to the botanist Diana Beresford-Kroeger, the xylem and phloem that transport vital fluids inside trees may be especially long in old-growth forests, and resonate in ways that birds find attractive, encouraging them to nest.

Plants reverberate both literally and figuratively. Literally, there is a vine in the Cuban rainforest that has evolved bowl-shaped leaves that act as sound reflectors. The leaves help echolocating bats home in on the vine's flowers twice as fast as they do those of other plants, and in return for a drink of nectar the bats pollinate the vine. Figuratively, the poet Bashō imagines that he hears the sound of temple bells continuing in the flowers after the bells themselves have stopped.

In the poem 'A Tree Telling of Orpheus', Denise Levertov imagines a tree that is moved to dance by the musician's song.

But can plants actually hear? Charles Darwin tried to find out by playing a bassoon to a mimosa. He wondered if it might respond by closing its leaves, just as it does when gently touched. The mimosa did not move, however, and Darwin concluded that he had been engaged in a 'fool's experiment'. His lack of success did not deter a mezzo-soprano named Dorothy Retallack, who in the 1960s showed – to her own satisfaction at least – that plants grow better when exposed to Bach but shrivel under the influence of Jimi Hendrix and Led Zeppelin.

Retallack's method was deeply flawed, but more refined tests in the 1970s did at first seem to suggest that corn germinates more rapidly when exposed to music of any kind, whether Mozart or Meatloaf, than to silence. But this finding turned out to be wrong too. It was actually warmth from the loudspeakers that was making the difference.

By the early twenty-first century, some researchers were arguing with confidence that members of the plant kingdom are insensitive to sound. Yes, many trees and other plants are exquisitely sensitive to signals passed by others through mycelial networks in the soil, and in other ways. But what advantage could there be for plants in detecting sounds, and how could they possibly do so without brains, or even nervous systems? It seemed like the case was closed.

Except it wasn't. It turns out that plants such as the beach suncup, or beach evening primrose, do hear the sounds of animal pollinators. Using its flowers to magnify and concentrate the sound like an old-fashioned ear trumpet, the suncup reacts to a bee's buzzing wings by increasing the concentration of sugar in its nectar. It does this in less than three minutes – fast enough to make a difference to a bee that has been exploring nearby before it decides to touch down. Even if the bee flies away too quickly, the plant is ready to better entice the next one.

Researchers have also increasingly realised that some plants are remarkably sensitive to vibrations and can discern what is causing them. Thale cress can detect vibrations of less than one ten-thousandth of an inch, or 0.00254 millimetres, in their leaves caused by insects chewing on them, and release a repellent chemical in response. The same plants, exposed to other vibrations caused by the wind or different insects, do not produce more of the chemical. Roots of the common pea plant, meanwhile, can locate water by the vibrations that the water makes when moving inside a pipe even though the soil immediately around the pipe is no wetter than the surrounding soil. The roots then start to grow towards the pipe.

Darwin's original experiment may have failed but, as was so often the case with him, his first intuition was not far off track. When he suggested in his penultimate book, *The Power of Movement in Plants*, that root tips act like the brains of simple animals, he was right. We still understand little as to *how* some plants process sound and organise their reactions to it, but we can add to Shakespeare's figurative 'tongues in trees and books in the running brooks' a real world in which some flowers actually hear.

It is also becoming increasingly easy for us humans to listen to the sounds that plants make as they grow and respond to their environment. Microphones placed on the trunk of a tree can capture the sound of water and nutrients channelling through cells: a kind of delicious woody drinking noise. 'The sounds of plants are much more fascinating, and challenging to our perception of them, than we have yet to realise,' says Jez riley French. As technologies for listening to them become more accessible, he suggests, there is an opportunity to learn to respect and represent the ecologies of which they are a part with more equity; we can learn that ecosystems that thrive depend on roots and soil, not just on eye-catching leaves and flowers.

Insect

Insects are not like you and me. They have more ears. And in some very strange places. Some butterflies and moths have them on their mouth parts. Others have them at the base of their wings, rather as if we had ears in our armpits. One group of butterflies known as the satyrines channel sound to these tiny ears thanks to swollen veins on their wings, making their means of flight the equivalent of our outer ear flaps, or pinnae. Crickets have organs of hearing on their front legs. Flies listen with sensors on their antennae. Praying mantises hear through a single ear in the centre of their thoraxes. Grasshoppers detect sound through membranes on their abdomens.

The almost endless and surreal variety in insect anatomy goes far beyond how they hear. There is, for instance, a butterfly with an 'eye' – strictly, an 'extraocular photoreceptor' – on its penis. But the fact that so many different kinds of ear have evolved, and have done so independently at least nineteen times in different groups of insects, shows just how important sound is for many (though by no means all) of them in the struggle to survive and reproduce.

Arthropods – the phylum of invertebrates with exoskeletons and segmented bodies that includes insects – emerged from the sea on to the land more than 450 million years ago. Some came to lay eggs in the sand of Ordovician beaches, just as horse-shoe crabs do today. Others came to find food, such as algae, simple plants, worms and other arthropods. The first insects (arthropods with six legs) evolved about 400 million years ago in terrestrial environments that were already teeming with centi-pedes and millipedes as well as with predatory scorpions and spiders (arthropods with eight legs). Most or all of these crea-tures could probably detect vibration in the soil or vegetation through sensors in their legs, so there were strong incentives to keep as quiet as possible and to listen out for danger.

But there were also good reasons in some cases to make sounds and vibrations deliberately. A sudden buzz can startle a predator and buy time to escape, and this capability – the same principle as a hand buzzer practical-joke device, but to a serious end – probably evolved early on. It's still found in spiders, milli-pedes, crickets, beetles and woodlice today. Further, advertising your presence with sound can be helpful when you're trying to find a mate or scare off rivals.

Insects probably created the Earth's first terrestrial songs – or, in the jargon, 'Palaeozoic intersex calling interactions'. Exactly how and when they did so is unknown, but there were probably at least two different origins. The first was precisely to take advantage of vibrational sensing – that ability to feel a buzz, rattle or other movement transmitted through vege-tation or the ground. Some of the results can be heard today among treehoppers, a family of small sap-sucking insects that are distinctive for the enlarged structures on their backs that often resemble thorns but which can also take on fantastical colours and shapes, like psychedelic *Pickelhauben*, helicopter rotor blades and crescent moons. By rapidly contracting the muscles in their abdomens, treehoppers make vibrations that pass through the plant they are standing on and up the legs of other treehoppers. These vibrations are inaudible to humans, but can easily be converted into sound by electronic means.

A library of recordings compiled by the biologist Reginald Cocroft contains songs by different species of treehopper that sound, variously, like a scratchy didgeridoo, a hooting monkey combined with mechanical clicks, and the warning noise that a truck makes when it's reversing, combined with a drum. With surface vibrations, these tiny insects can make noises that seem as if they're coming from much larger bodies, like arthropod versions of the Wizard of Oz in the 1939 film, conjuring tremendous sounds from their leafy green booths. A treehopper can produce a mating call as low as that of an alligator, even though the alligator is many million times heavier.

A second origin for insect song is in an adaptation of body parts that had already evolved for the remarkable innovation of flight. In this pathway, singing came from winging. Studies that trace back the shared roots of different genes in divergent living species suggest that the first winged insects evolved between 400 million and 350 million years ago, while the earliest fossil evidence for modifications to these wings to broadcast sound dates to approximately 310 million years ago, in the late Carboniferous. Specialised zones on the wings of *Titanoptera*, a blackbird-sized insect related to modern grasshoppers, probably crepitated – that is, made a crackling noise – when brought together, and may even have produced flashes of light. Both would have helped the animals communicate.

The oldest uncontested evidence for stridulation – the scratchy-singing sound made by rubbing wings or other limbs together – comes from fossils laid down tens of millions of years later, in the Permian period. *Permostridulus*, a distant cousin of today's crickets, had a tiny thickened and raised vein on one of its wings. This would have scraped over the base of the other wing when they were rubbed together. The simple rasping sound was the start of a 300-million-year symphony.

In a more sophisticated version of what is essentially the same structure, modern crickets draw a nub on the left wing over ridges on the right, like a plectrum across a comb. They also amplify the sound with a drum-like 'window' of membrane in the wing. The shape of the file and window is different in each species of cricket, as is the rhythm, and they make a great variety of songs, from gentle chirping to high trills and whines that are above the range of human hearing. Some crickets amplify their song by chewing a hole in a leaf, sticking their heads through the hole and using the rest of the leaf as a megaphone. Other kinds of insects have evolved quite different ways to make sounds. Cicadas, for instance, have tymbals – corrugated structures on their abdomens which they vibrate rapidly over resonance chambers in the sides of their bodies to make some of the loudest sounds of any insect.

Permostridulus and other insects of the Permian period lived in a world where there may have been few other animal sounds beyond incidental squelches, plops, hisses, thumps and scrapes. Our own distant ancestors at that time would probably not have been able to hear their calls. Therapsids – the galumphing lizard-like creatures from which we are descended – were only sensitive to low-frequency sounds. Over time, however, some of our cousins shrank in size, grew fur and in some cases learned to fly, becoming formidable predators upon insects.

Bats evolved the ability to echolocate – that is, to emit loud high-pitched sounds and read the returning echoes in order to locate objects in space – around 52 million years ago. This enabled them to find and follow flying insects even in complete darkness. They gorged themselves, and flourished: one in every five species of mammal alive today is a bat. But insects evolved countermeasures. When a praying mantis hears a bat with the ear in its thorax, it tumbles like a fighter pilot who allows their plane to fall suddenly in order to evade a missile. Timing is crucial, but the mantises get away about 80 per cent of the time. In many moth species the ultrasonic clicks of a bat picked up by their tympanums trigger the flight muscles to twitch erratically, jerking the insect's trajectory in random directions which are harder for the bat to follow. Other moths have evolved even more ingenious defences. Tiger moths have bumps on their exoskeletons that buckle to make ultrasonic clicks which jam a bat's echolocation, and also signal that the moth is poisonous. Luna moths have tail-like appendages which spin and flap as they fly, creating acoustic decoys that confuse the bat. There are even 'stealth' moths with scales on their wings that work like acoustic tiles on an advanced fighter aircraft, soaking up around 80 per cent of the sound in the bat's clicks and making any return signal hard to detect.

There is at least one case where a butterfly has learned to use sound to exploit other insects. Caterpillars of the Alcon blue, a butterfly with spotted wings that is found across Europe and northern Asia, produce a sweet substance that attracts Myrmica ants and then mimic the song of their queen. The worker ants will adopt the caterpillar and carry it back to their nests, where it continues to sing, and they feed it in preference to their own larvae while it grows up to a hundred times its previous size: a monstrous invertebrate cuckoo.

The more we learn about the sound worlds of insects and other invertebrates, the more there is to be astonished by. It was recently found, for instance, that black widow spiders can tune vibration in their legs to different frequencies by changing their posture on their web. 'It's as if a human could focus on the colour red by squatting, or single out high-pitched sounds by going into downward dog (or downward spider),' observes Ed Yong. The wings of satyrine butterflies, meanwhile, come close to the performance of an ideal microphone, which accurately represents all surrounding sound without preferential amplification of certain pitches over others – something that humans find hard to achieve even with sophisticated technology.

In 1962 Rachel Carson warned of a silent spring – an absence of birdsong as birds died out in huge numbers because their principal food sources had been contaminated by the indiscriminate use of pesticides. 'Today,' writes the biologist Dave Goulson, 'she would weep to see how much worse it has become.' The problems that she highlighted are far more acute, in part (though not only) because new insecticides are thousands of times more toxic than they were a few decades ago. So it won't only be the birds who fall quiet. In Britain the number of flying insects fell by almost 60 per cent between 2004 and 2021. Other countries are reckoned to have seen similar declines. 'Few people seem to realise how devastating this is,' says Goulson. The loss of insects will, he says, profoundly affect human well-being because we need them to pollinate our crops, recycle dung, leaves and corpses, keep the soil healthy, control pests, and much more. It will also affect birds, fish, frogs and other animals who rely on insects for food, and wildflowers, which rely on them for pollination. If insects continue to become scarcer the living world as we know it will slowly grind to a halt, for it cannot function without them.

Bee

Are bees upset by echoes? It's not a question many of us might think to ask. But, finding this 'wild and fanciful assertion' in the writings of the Roman poet Virgil, the eighteenth-century parson and naturalist Gilbert White decided to investigate for himself. Holding a large speaking-trumpet close to the hives in his garden, he spoke 'with such exertion of voice as would have hailed a ship at a distance of a mile'. The bees, he reported, 'pursued their various employments undisturbed, and without shewing the least sensibility or resentment'.

That bees are indifferent to many sounds, and possibly deaf, was not news in White's time. Aristotle, who lived three centuries before Virgil, had said as much, and White knew that the natural philosophers of his own day maintained that the creatures 'are not furnished with any organs of hearing'. But White, a lifelong beekeeper, and an acute observer of the natural world, wondered if the philosophers might be missing something. Could bees *feel* the repercussions of sound – that is, vibration – even if they could not *hear* them? Sometimes there is more going on than meets the ear.

Humans have long exploited bees, and continue to do so. As well as keeping them and harvesting their honey, we have projected on to their colonies all kinds of ideas of how our societies should, or should not, be organised. But at the same time, we have also marvelled at them as beings in their own right. 'Some have claimed for bees a share of divine intelligence and a draught of the springs of heaven,' wrote Virgil in the same Eclogue in which he noted their sensitivity to echoes. Pliny, born a century later, suggested in his *Natural History* that their honey might be the saliva of stars, or the perspiration of the sky.

And the science of bees is no less astonishing than the stories people have made up about them. For a start, bees are enormously diverse. In addition to nine species of honeybee and 250 or so of bumble, there are more than 20,000 other kinds, and they live in gloriously various habitats. Amid the organ-pipe cactuses, teddy-bear chollas and Boojum trees of the Sonoran Desert in North America, for instance, lives *Perdita minima*. At less than two millimetres across it is the world's smallest bee, and it thrives there alongside a carpenter bee which is more than twenty times its size, as well as scores of other kinds.

Bees are flexible and adaptive learners rather than preprogrammed robots. They navigate and remember complex and unfamiliar terrain, and share what they learn with others – but this is just the beginning. Honeybees, for instance, can recognise individual human faces. Bumblebees will deliberately damage leaves on some plants because this reduces the time the plants take to flower and produce pollen. Some bees can also solve puzzles they would never encounter in a natural setting, such as learning to access an artificial flower visible under a transparent plate by pulling on a string. Recent research by H. Samadi Galpayage Dona suggests that bumblebees will engage in behaviours, such as rolling wooden balls in the lab, that fulfil all the criteria for animal play. They do all this with a brain about the size of a grain of sand, and fewer than a million neurons – about 0.001 per cent of the 86 billion in a human brain. Each neuron in a bee's brain has as many branches as an oak tree, and makes synaptic connections with thousands of others.

Would creatures so capable really not make use of sound? Following in Gilbert White's footsteps, the writer Maurice Maeterlinck noted in 1901 that bees were not in the least disturbed by the noises that he, as a beekeeper, made near their hives, but suggested they were merely ignoring him. Could it be, he asked, 'that we on our side only hear a fractional part of the sounds that the bees produce, and that they have many harmonies to which our ears are not attuned?'.

Maeterlinck's inference was corroborated by the pioneering biologist Charles Henry Turner, who showed that bees can hear and can distinguish pitch. Further evidence came from observation of what is known as the waggle dance. In this behaviour, which was first studied in detail in the 1920s by the ethologist Karl von Frisch, a successful forager bee performs a shuffling figure-of-eight move inside the hive in order to share information about food and water sources or potential new nest sites. As the bee waggles it releases compounds called alkanes and alkenes, and creates a distinctive electrical field. Both of these are probably part of communication, but researchers now think that surface and airborne vibrations created by the waggle play a role too. 'During the waggle phase, the wing-beats of the dancer produce a train of vibration pulses that pass from the tail end of the dancer to a follower bee,' writes the researcher Hiroyuki Ai and his colleagues. The followers detect the airborne vibrations, or near field sounds, in what is known as Johnston's organ on their antennae, and process the information in their primary auditory centres – a part of their nervous system visible to twenty-first-century neuroscientists but not to eighteenth-century natural philosophers. It has also been found that when bees bump into each other they go 'whoop!'. At first, researchers thought that this was a signal to the other bee to stop, but it now appears they are merely surprised.

There is also some evidence that honeybees can detect sound over a greater distance outside the hive. Manufacturers of high-end audio systems have recorded the hum of a hive and then replayed it some metres away. Bees from the hive appear to approach the speaker replaying the recorded sound in significant numbers.

Further, bumblebees use vibration as a tool. There's a good chance you've heard them doing this if you've spent time on a summer day in a place with abundant flowers. Every now and then, the lazy, deep buzz of a bumblebee becomes a series of higher-pitched bursts, a little like the sound of a tiny drill going into wood, or an agitated bluebottle caught in a spider's web. Some people wonder if the bee is distressed, but this is not what is happening. Many plants transfer pollen from their flowers to pollinating insects simply by allowing them to bump into their anthers, but about one in twelve plant species, including many that are valuable food sources for humans, have tube-like, 'poricidal' anthers where the pollen sits inside and can only escape through a small pore at the tip. The high, intense buzzing of the bumblebee, made by contracting its flight muscles and generating a g-force of up to thirty, causes the pollen to pour out. It's known as buzz pollination, or sonication.

The story goes that not long after the end of World War One the movie mogul Samuel Goldwyn recruited a stable of eminent authors in a bid to elevate the tone of his films. One of them was Maurice Maeterlinck, who by this time had won the Nobel Prize in Literature, and the great man set to work on a screenplay. When it was translated, Goldwyn began to read, at first with puzzlement and then with increasing dismay until he could no longer contain himself. 'My God!' he stormed, 'The hero is a *bee!*'

Maeterlinck's Hollywood career ended then and there, but his love for bees continued to inspire others, and his intuition with regard to sound have proven correct. Bees really do have 'harmonies to which our ears are not attuned,' and there may be more to discover.

There is wonder in learning about sounds in the living world that are beyond our hearing, and it is often fascinating when these are made audible by such means as electronic amplification. But it can be enough, or better, sometimes, simply to pay closer attention to what is already available to us with our unmediated senses. 'Attentiveness alone can rival the most powerful magnifying lens,' writes the biologist Robin Wall Kimmerer with regard to the study of mosses. And so, too, with bees.

It is worthwhile, at least once in your life, to get your ears as close as you can, safely, to an active beehive. 'Between us and heaven or hell is only life, which is the frailest thing in the world,' wrote Blaise Pascal. But in the intensity and spectrum of the sound, where thousands of bees organise and cooperate as a collective intelligence, life can be heard bursting with power and possibility far beyond our own brief span, if only we allow.

Frog

What a piece of work is a frog. How noble in breathing, which it does through its skin. How express is its tongue, ten times softer than a human's and coated with saliva a hundred times stickier. How admirable in swallowing, during which it squeezes its eyeballs down inside its skull to help push food down its throat. For apprehension, a rod-based colour vision system enables it to see colours even when it is so dark that humans see nothing at all.

The first true frogs are thought to have evolved in the Jurassic almost 200 million years ago. *Beelzebufo*, an early member of the clan, was large enough to chomp baby dinosaurs. Today, more than seven thousand species live on every continent except Antarctica. They range from *Paedophryne amauensis*, the world's tiniest vertebrate, which is smaller than the eraser on the end of a pencil, to the Goliath frog, which is big enough to go twelve rounds against a Toy Poodle. In form and movement, there is almost no end to their facility. Some glide through the air in the rainforests of Borneo and Sumatra on parachute-feet, while in Lake Titicaca an entity so wrinkled that it is sometimes called the scrotum frog never leaves the water. Australian burrowing frogs survive for years inside mud. In Vietnam, mossy frogs blend almost perfectly with riverside stones. The poison frogs of Central and South America sport high-vis mottles and stripes of gold, blue, red, orange and black.

Sound is a big deal on planet frog. For the male of a species that advertises itself to females by calling, the deeper the call the more attractive you are. And frogs can't fake this: 'expressive size symbolism' means that the bigger they are, the deeper their call: for them, truth is beauty, and beauty truth. 'If you're a male and you don't want a fight,' says the ecologist Peter Warshall in his lecture on 2 billion years of animal sounds, 'you go over to another who's already mating with a female, and tap the guy on the back. If his voice is deeper than yours you move on. If not, you swap places.' Frogs have been doing this a very long time so they know the routine. Sometimes, if you're a male frog, you have to tap a lot of others on the back before you get to kiss your princess.

Female frogs are also sonic wonders. In the warm, wet places that many kinds of frogs like to live, males of dozens of different species croak at the same time, and this creates what is known as the cocktail party problem: the challenge of hearing the right voice in the ruckus. To overcome it, the females make sounds in their lungs at exactly the frequencies of the calls of other species. They transmit these to their own eardrums and – on the same principle as noise-cancelling headphones – muffle them out. Thanks to this, the voices of males of their own species stand out clearly, and a rising generation, perfect in every particular, will soon crawl out of the water to begin the game anew.

For some, frogs are a source of sounds and sweet airs that give delight and hurt not. 'The steadily increasing sound of toads and frogs along the river with each successive warmer night . . . is the first earth-song,' wrote Henry David Thoreau in May 1860. It is 'as if the very meads at last burst into meadowy song'. For the musician Hermeto Pascoal, frogs croaking on a rainy day offer an invitation to duet, and he joins them on the piccolo.

But frogs do not always please humans or our objects of worship. In the eponymous play by Aristophanes first performed in 405 BCE, their croaking infuriates the god Dionysus. And in 2014 insufferably loud frog sex nearly drove the actress and singer Ariana Grande to distraction. Fortunately for humanity, Grande found an ocean sounds app on her phone to block them out.

Frogs are under multiple assaults from pollution and other disturbance caused by humans. But there may be hope for their continued existence, not least because some frogs can mime. Near roaring waterfalls, where it is too noisy to hear each other, they have learned to simply wave hello to attract mates.

Bat

That bats echolocate, finding their way and catching insects by listening to the echoes of their own calls, is such a commonplace today that many of us seldom give it further thought. This is a pity, because the more one considers it the more amazing it is.

Start with the history. Surgically removing the eyeballs of a bat was just another day in the life of Lazzaro Spallanzani. In 1786, this priest, biologist and man of parts had been the first to fertilise frogs *in vitro* and artificially inseminate a dog, and now, in 1793, he was trying to work out how small furry mammals were able to fly without mishap in a space so dark that even a barn owl came to grief. When a bat he had blinded continued to fly just as well as it had done before, Spallanzani concluded that it must be finding its way by some other means than sight. Searching for an alternative, he plugged the ears of unblinded bats with wax and other substances, and when the deafened animals careened into walls he decided they must somehow be finding their way by listening – though exactly how he could not tell.

The answer did not begin to come into view until more than a hundred years later, at a time when advances in technology were expanding the realm of the conceivable and measurable. In May 1912, one month after the sinking of the *Titanic*, the meteorologist Lewis Fry Richardson filed a patent for an underwater echo-ranging device, or what we would now call sonar; and that same year Hiram Maxim, who is better known as the inventor of the automatic machine gun, proposed that bats navigate by listening to echoes of sounds lower than the limit of human hearing. Eight years later, the physiologist Hamilton Hartridge suggested that they actually make sounds higher than human hearing, and this was confirmed by Donald Griffith, a young biologist, who began to study the creatures in the 1930s with a newfangled device that could detect those high-frequency sounds. In 1944 he coined the term echolocation. It is, then, only a single human lifetime since Griffith (who went on to champion the unfashionable idea that many animals can think and reason) and his co-researchers proved the hitherto almost unimaginable: that bats 'see' with sound.

There's an excellent account of the *Umwelt*, or sensory universe, of bats in *An Immense World*, a book about animal perception by Ed Yong. Here are a few key points. First, bats may seem almost silent to us, but they are in reality almost unbelievably loud. A big brown bat, which weighs just fifteen to twenty-six grams, or about as much as three to five credit cards, screams at 138 decibels, the same volume as a jet engine. Even 'whispering' bats holler at up to 110 decibels, or about as loud as a chainsaw. Humans don't hear them because the frequencies at which bats call – at up to 200,000 hertz – are far higher than anything our ears can process, but the noise is real enough.

The reason bats scream so loudly is that at very high frequencies sounds are quickly absorbed by the air, and don't travel far. To compensate, the animals tend to funnel the energy of their calls into a cone which extends from their head like a narrow beam of a headlight. With this they can find prey such as moths or other insects as much as five to eight metres away. And to avoid deafening themselves they contract the muscles of their middle ears to desensitise their hearing in exact synchrony with each call and relax them in time to hear each echo.

To avoid obstacles and locate their prey, bats need to rapidly update what would otherwise be a static snapshot in sound with fresh calls and echoes. They do this with vocal muscles that, in the 'terminal buzz' when a bat is homing in on a diving insect, contract at up to 200 times a second, the fastest movement of any mammalian muscle. This keeps each call to a few milliseconds so that even during the rapid terminal buzz there is no overlap between call and echo that might confuse the bat.

The precision of this see-hearing in bats is astonishing. Researchers have found that the animals can detect differences in echo delay of one to two millionths of a second – the time it takes sound to travel less than a millimetre. This is better acuity than the human eye over a comparable distance. Further, some species echolocate with calls over a range of pitches spanning more than an octave: one might imagine the world's highest and wildest slide whistle. The lower frequencies over this range tell the bat about larger features of its target, while higher frequencies describe fine detail. Somehow, the bat analyses all this information, and more, in a fraction of a second to build a detailed acoustic image of its prey and its direction of travel. Researchers have recently discovered that some bats also make much lower noises to greet each other. They are tiny death metal growlers and Tuvan throat singers on the wing.

The biologist Leslie Orgel once said that 'evolution is cleverer than you are'. He didn't mean this to be taken literally, of course: evolution doesn't have a mind in the sense we normally use the word. The point, rather, is that evolutionary processes give rise to forms and capabilities that few if any of us would have been able to think up. Or, as I like to put it, evolution is not only smarter than you; it has a stranger imagination.

Imagination – defined by the philosopher Stephen Asma as the ability to decouple the mind from the immediate flow of perception and run simulations of counterfactual virtual realities – is (as far as we know) much more developed in human beings than in any other animal. It is, perhaps, our most remarkable capacity. But it is one that we can forget or abuse, and when we do so we get stuck or enter a downward spiral.

So one of the things I recommend to rekindle the imagination is to take a good look at the faces of bats. The gigantic strangeness and the surprise of evolution is writ small in their astonishing and varied faces – some of them more bizarre than any mask, gargoyle, demon or alien dreamed up by humans. Not all are adapted for echolocation. Many species of bat find their way, their mates and their food by sight and smell, but each one is an ingenious (if weird-looking) answer to the challenges of survival.

Some of that diversity is celebrated in an illustration by Ernst Haeckel in his *Artforms of Nature*, published in 1899. In this extraordinary compendium of images, which include jellyfish, sponges, radiolaria and other creatures as well as fungi and plants, Haeckel arranges and accentuates the bat's features for dramatic effect but does not misrepresent them. The lesser long-eared bat, *Nyctophilus geoffroyi*, is a frowning stoat-god with elephant ears. The Antillean ghost-faced bat, *Mormoops blainvillii*, is a cross between a lion with walrus moustaches, a jelly-ear fungus and the mythical Green Man. The greater horseshoe bat, *Rhinolophus equinus*, is a gimlet-eyed Totoro bedecked with curlicues.

And we are only getting started. Haeckel shows just fifteen different kinds of bats, but the world contains at least 1,400 – about one in five of all mammal species. Among my many favourites beyond Haeckel's page is the hammer-headed bat, *Hypsignathus monstrosus*, which has a tubular snout a little like that of a Saiga antelope but ending in a warty orchid crossed with a *memento mori*. The greater bulldog bat, *Noctilio leporinus*, meanwhile, has gimlet eyes, bulldog jowls and strangely downwards-pointing ears – these last an adaptation with which it senses telltale ripples in the surface of water caused by the underwater movements of its fishy prey. These decidedly weird heads and detection apparatuses are actualisations of just a few of the many ways in which a creature can find a way to flit briefly through existence on this quickly turning planet.

Elephant

Most humans can sing over a range of about two octaves, although Freddie Mercury could manage three or four, and the performer Tim Storms holds the record at nearly ten. For African savannah elephants, however, a ten-octave vocal range is normal. Adults can rumble at far below the threshold of human hearing, and even calves make noises as deep as those of the lowest pipes on a church organ. In the mid-range, the elephants snort and trumpet, and they can also vocalise considerably higher than the top note on a piano. African forest elephants and Asian elephants, the other surviving species in the elephant family alongside their savannah cousins, have comparable abilities.

Like humans, elephants hear through the air, acoustically, but they do so with much more acuity. In favourable conditions, which are typically around dusk in calm, cloudless weather, they can detect the calls of others of their kind over ten kilometres (six miles) away. But they can do more than this, for elephants also hear seismically – with their feet.

To sense vibrations through the ground, an elephant will lean forwards to place more weight on its front feet, and stand stock-still as it allows the vibrations to pass up through its bones to the cavities of its middle ears. At the same time, sphincter-like muscles contract and close the ear canals, dampening acoustic signals and enhancing seismic detection. Pressure building up in a sealed air canal creates a closed tube which enhances bone conduction. As well as freezing in place, an elephant will sometimes lift one foot off the ground and replace it, or reposition itself with respect to the general direction of the vibration in order to get a more precise bearing, rather as we may turn our heads to tune into the direction of a sound.

The bones in an elephant's feet do not press flat along the ground as they do in human feet, but are supported at the heel by a thick pad of flesh which tilts them up and forwards, much as the wedges on high-heeled shoes tilt up the bones in a human foot. The wedge that is the rear part of the elephant's flat-bottomed sole is filled with fat and cartilage that maximises the transmission of vibrations to sensitive mechanoreceptors, or touch cells, like the ones we have in our fingertips and genitals. And thanks to these cells, which are called Meissner and Pacinian corpuscles, elephants can identify very small changes in frequency – fractions of a semitone – in the grunts and noises that other elephants send through the ground as alarm signals or just to let them know they are around. It is thanks to this sensitivity that adult males and females, who tend to live far apart most of the time, find each other during the brief window – lasting just four days, and occurring once every four years – that females are in oestrus.

Vibration travels further and faster through the ground than sound does through the air, and elephants can detect things seismically that are considerably further away than those they hear acoustically. They can, for instance, sense the thump-thump of helicopter blades more than 130 kilometres (eighty-one miles) away, identify it as potential danger, and run off in the opposite direction. It is thought they can feel rain thrumming on the ground as much as 240 kilometres (149 miles) away.

Seismic detection can be so effective that one of the elephant's cousins has given up almost completely on listening to sound waves. Save for the gorgeously iridescent sheen of its fur, the tennis-ball-sized golden mole looks almost exactly like other moles but it is actually more closely related to elephants than it is to them. (Both are members of the Afrotheria, a clade of mammals which evolved from a common ancestor in Africa millions of years ago when it was an island continent.) Shunning sound waves, the golden mole has evolved a malleus bone in its middle ear so large that when it presses its head into the Namibian sand to listen for the vibrations emanating from distant patches of grass swaying gently in the wind where it may find food, the malleus stays still while the other ear bones and indeed most of the mole vibrate around it.

Manatees and dugongs are even more closely related to elephants than are golden moles, and they, like elephants, have pads of special acoustic fat in their skulls which help to conduct sound to where it is needed. They too may be described as hypersensory. Where elephants have a highly dextrous, sensitive and strong trunk, capable of picking a ripe peach without bruising it at one moment and knocking over a brick wall the next, manatees have an oral disk: muscular, prehensile 'lips' that can handle and explore objects as delicately or firmly as a human hand. I want one. Dugongs are noted for their exquisitely well-developed sense of touch, as well as sensitive hearing. According to the novelist and artist Jonathan Ledgard, humans long placed their inner-ear bones in sacred mounds as a magical device, endowing their guardians with superpowers.

Researchers are only beginning to understand some of the ways in which elephants use sound and seismic waves, not to speak of many of their other behaviours. If, as was shown in 2014, these animals can determine the ethnicity, gender and age of humans from acoustic cues in our voices alone, what else may they be able to do? It is not too late to learn more about the superpowers of the three surviving species of elephant but, endangered or critically endangered as they are, it is getting there.

The Thousand-mile
Song of the Whale

The seas are to sound as space is to light. Sound travels more than four times as fast underwater as in air and can carry for vast distances. Light, by contrast, actually passes more slowly through water than it does through air. Sunlight is quenched almost completely by about 200 metres down, while sound continues unimpeded thousands of metres downwards and for almost unlimited distances horizontally.

Tens of millions of years ago, the ancestors of today's whales and dolphins – descendants of wolf-like animals that hunted at the edge of the sea – evolved to take advantage of this. They learned to use sound to track their prey to great depths, and communicate with their companions over long distances. As a consequence, a whale's sense of the world, self and others is mediated far more by sound than it is by sight.

Over time these ancestors evolved into two distinct groups: those with teeth and those with baleen. Toothed whales, or *Odontoceti*, such as dolphins, orcas, narwhals and beaked whales, use their teeth to catch fish and other prey. Baleen whales, or *Mysticeti*, which include bowheads, right whales, grey and blue whales, filter plankton from seawater through comb-like plates – baleen – in their mouths. The two groups inhabit very different sound worlds. Toothed whales use echolocation to hunt, emitting high-pitched clicks which bounce off their prey and return to their ears. They also communicate with each other with clicks and whistles. Baleen whales do not, as far as we know, use sound to hunt, and sing to each other in continuous, and often lower, tones.

Before humans filled the seas with human noise, the songs of baleen whales would have carried across entire ocean basins and even into other oceans through the deep sound channel, a kind of lens formed by pressure and temperature gradients hundreds of metres beneath the surface that can transmit sounds over thousands of kilometres. If the dream of the composer R. Murray Schafer of the world itself as a macro-musical composition has ever been realised, it is in substantial part thanks to these whales.

The first human encounters with whales and dolphins were probably chance sightings from shore, but early seafarers would have heard them too as their songs penetrated the hulls of small wooden vessels. It may be that Mediterranean sailors of antiquity interpreted such sounds as the calls of the sirens who lured Odysseus. In some places, seafarers learned early to track whales by ear, as well as by watching for them to blow at the surface, listening for their cries and clicks underwater by dipping a paddle or spear in the water and placing the other end against their skull.

Humans have been hunting whales for a long time. Petroglyphs between four and eight thousand years old found at Bangudae in what is now Korea show grey, sperm, humpback and right whales in large numbers swimming and jumping freely, but also being killed. Indigenous peoples such as the Iñupiat in what is now Alaska have been interacting with bowhead whales for many generations, hunting them on a small scale but also engaging in a respectful relationship in which the bowheads are regarded as beings of great spiritual worth. Systematic, large-scale hunting probably began with the Basques, who started to hunt right whales in the Bay of Biscay as early as the eleventh century, and who by the 1500s were chasing rights and bowheads off Newfoundland. Throwing double-flued harpoons fixed to long braided ropes from small boats, they killed thousands in the first centuries of what was to become global business. Over the course of the nineteenth century, whalers from Britain, the United States and elsewhere killed sperm whales and other species by the hundreds of thousands. In the twentieth century, factory ships from more than a dozen nations slaughtered millions, and almost eliminated many species. About 99.85 per cent of blue whales, the largest of all, were killed at this time.

In 1966, the biologist Roger Payne, who had been studying echolocation in owls and bats, heard recordings of whale song made by Frank Watlington, a US Navy engineer at a secret station listening for Russian submarines off the coast of Bermuda. This first encounter, Payne said later, was like hearing the size of the ocean. 'It was as if I had walked into a dark cave to hear wave after wave of echoes cascading back from the darkness beyond . . . That's what whales do, give the ocean its voice.' And as Payne as his co-researchers started to analyse these strange sounds they became more and more astonished.

Humpbacks go *ooo-oop* and *pwfwhaa!* They go *eeeouu* and *yeeeee*. They make noises like creaking doors, like mopeds rattling in low gear and prolonged farts à la *basso profondo*. Some of these sounds – including the fart sound – are towards the bottom of what humans can hear, at twenty hertz or even lower, while others are towards the top, at 20,000 hertz and beyond. Individual sounds last a few seconds, and may become louder or softer as they are sung. At first hearing, a sequence of them can seem random – a rattlebag of sonic curiosities – but they are actually highly organised.

Roger Payne and his wife Katy, together with Scott McVay and his wife Hella, made the first steps in deciphering the structure of these songs. Using a device for printing images of sounds, they created sonograms of what they had recorded. The technology of the late 1960s was slow and primitive by today's standards: each ten seconds of sound took an hour to print. But, as they gradually assembled these fragments, Hella, a mathematician, began to see structure and organisation. Tracing the sonograms by hand to render the songs with less extraneous noise, the four devised a system of notation to show this. The smallest building block of the song structure, which they called a unit, was any one of those diverse vocalisations lasting a few seconds each. Repeated units were grouped together in what they called phrases, which a whale would typically sing for twenty to forty seconds. A sequence of phrases made of different units made a theme, and a song was a sequence of themes, typically lasting anything from about seven to thirty minutes. In what they called a session, a whale would repeat the same song many times, sometimes for a whole day. Presenting their work in the journal *Science*, Roger Payne and Scott McVay wrote that in whale song 'beautiful and varied sounds' were repeated 'with considerable precision', sometimes for many hours. They had shown that, far from being a series of random cries, the sounds are the deliberate creations of complex minds. Many researchers now agree that whale song meets a reasonable definition of music.

In the following decades, researchers found that the songs that individual whales sing and copy from each other in one part of the ocean are different from those of the same species in another. Katy Payne, her colleague Linda Guinee and others also showed that the songs change over time, with individuals introducing variations to gradually create new songs that can eventually spread from whale to whale across entire oceans. They even showed that the songs sometimes rhyme, with the end of different phrases mirroring each other. Exactly why whales sing these songs, however, remains a mystery. Males do not seem to be performing for females, for instance. It may be that they are all just trying to impress each other. Perhaps they are just playing.

Roger Payne had been distressed by cruelty shown by humans towards marine mammals and, recognising the potential of the songs in a culture that was increasingly becoming aware of a history of mass slaughter, put together an album of recordings. *Songs of the Humpback Whale* was released in August 1970, just a few months after *Let It Be* by The Beatles. It was an almost instant hit and more than 10 million copies were eventually sold or distributed during a wave of heightened concern that gave rise to the modern environmental movement. In 1972 the United Nations declared a ten-year global ban on commercial whaling. Only a few countries ignored the ban, and in the following decades many species of whale began slowly to recover.

Recordings of whale songs have now become so familiar that it can be easy to forget how wondrous and strange they are when heard for the first time. One way to hear them afresh may be to start by listening to a human response rather than the recordings themselves. George Crumb's *Vox Balaenae* (Voice of the Whale), which was first performed in 1971, is scored for flute, cello, prepared piano and crotales (a percussion instrument consisting of small tuned discs also known as antique cymbals). The scope of the work is indicated by the titles of its three sections: 'Vocalise (. . . for the beginning of time)', 'Variations on Sea-Time' and 'Sea-Nocturne (. . . for the end of time)'. In the opening section a flute-player sing-talks at the same time as she plays – a technique that in various forms may date back to the most ancient playing of flute-like instruments. The strings of the piano, meanwhile, are played like a harp, creating a sense of space and depth. The second section begins with a whale-like sound played by the cello high in its register, answered with rumbles and vibrations from the piano. This is music not just for the age of whales but for the entire span of life on Earth, from the Archaeozoic to the Cenozoic era, in which whales evolved and in which we still live. In the final section, flute and cello produce an ethereal whistling that returns to the opening melodies above a lyrical piano, to be joined in turn by crystalline chimes on the crotales. Crumb said that in this conclusion he wanted to suggest 'a larger rhythm of nature, and a sense of suspension in time . . . serene, pure, transfigured'. *Vox Balaenae* has been described as an oceanic equivalent of Olivier Messiaen's music inspired by birdsong, but it is perhaps at least as much a counterpart to Mahler's *The Song of the Earth*.

Improvements in sound recording technology and more time in the field have enabled humans to hear more of, and in, whale song than ever before. The 2015 album *New Songs of the Humpback Whale*, which contains recordings by the musician David Rothenberg and others, reproduces the whales' voices with much greater clarity than on Payne's original LP. Listening to the newer album it's as if we are right there, next to the whales, rather than hearing them distantly through a column of water. Without the echoes and extraneous noise in the 1970 album there is, arguably, a loss of aura, but there is a gain in information and immediacy.

Other innovations can also help us appreciate whale song. Rothenberg and the designer Michael Deal have developed a system of graphical notation in which each discrete sound unit is assigned its own shape and colour. The shapes – which are like a cross between strange clouds and the neumes created in the tenth century to represent Gregorian chant – are stacked and sequenced along bass and treble clefs so that the range of pitches they represent is readily apparent, while their different colours help the human eye to distinguish between groups of them.

*

We love to listen to whales and dolphins, but what do they think of the noises we make? A Greek myth tells how dolphins saved the musician Arion from drowning after hearing his songs, and there have surely been many attempts in real life to speak or sing to them; but the earliest documented instance of a human playing music deliberately to cetaceans may have been a night in July 1845, when the cellist Lisa Cristiani performed Bach for a whale that came close to, and at one point passed directly under, the ship on which she was sailing in the North Pacific. 'From this moment,' she wrote, 'it became the received wisdom among our company that whales are purveyors of the finest tastes.' (An extraordinarily adventurous woman, Cristiani was on her way to tour Kamchatka with her Stradivarius when she died of cholera at the age of twenty-five.)

More recently, David Rothenberg has tried playing jazz with whales. For the recordings collected on the 2008 album *Whale Music* he stood in a boat and, through earphones connected to an underwater hydrophone, listened to the sounds of humpbacks, orcas, belugas and other whales. As the animals sang he responded by improvising on the clarinet, and the sound of the instrument was relayed in turn to the whales through an underwater loudspeaker. 'It seems to me as if at least one whale is reacting to me, changing his pitches and rhythms in response to mine,' Rothenberg wrote. 'This may be wishful thinking, but ... if the whales are able to change their songs over just a few weeks, one would expect them to be able to immediately respond to a whale-esque clarinet sound as well.' The tracks on *Whale Music*, in which the sounds of whale and clarinet are sometimes accompanied by guitar, violin and electronics mixed in afterwards, are striking and often beautiful, but it is hard to know how far, if at all, the whales are actually listening to Rothenberg and responding in some kind of antiphony, and whether there is any kind of pleasure for them in this encounter with a human. 'Do human voices sound as ethereal to the whale as whale voices sound to us,' asks the author Rebecca Giggs in her 2020 book *Fathoms: The World in the Whale*, 'or do we scratch and irritate the whale, a pin in the ear?' Sara Niksic, a biologist and musician who has produced two albums which combine whale song with psybient, a genre that mixes psychedelic trance and ambient, suggests that whales and humans, who both innovate and vary their songs endlessly, have more in common than many humans realise. But one thing is reasonably sure: whales attend on timescales that humans often find hard to match. A call made by a humpback near Bermuda would take twenty minutes to reach a humpback swimming off the coast of Nova Scotia, observes the science writer Elizabeth Kolbert. If that whale in Canadian waters answers immediately, it will be forty minutes before the Bermuda whale hears back. Kolbert quotes the bioaccoustian Christopher W. Clark: to imagine what it's like to be a whale, 'you have to stretch your thinking to completely different levels of dimension'.

Whether whales love or loathe human music, a record of some of the terrible hurt that we have inflicted on them in recent centuries can be found in their earwax. Fixed in that wax are stress hormones (as well as pollutants and other substances) that coursed through the whales' bodies while they were alive and the wax was forming. By examining plugs of it which have been extracted from the ears of dead animals, researchers can read their life histories in the dark bands laid down each year, rather as one can read a tree from its rings or a glacier from an ice core. Combining the readings from many individuals, it has been possible to compile a 150-year chronicle of whale stress across the world, and show that it matches almost perfectly with data on whaling intensity over the same period.

Today, commercial whaling has largely ceased, and many whale populations are recovering. But humans continue to hurt and kill whales in other ways. Fishing nets, whether in active use or as 'ghost gear' that has been left to drift, entangle many. The load of man-made pollutants may have already rendered some populations functionally extinct, and they may soon be gone altogether. And over the next few decades, whales from the poles to the tropics are likely to face multiple new sources of stress including higher levels of pollution and depletion of their food supply as a consequence of commercial fishing, and other factors. It may still be that, as the philosopher Amia Srinivasan writes, 'future historians will have the task of explaining how our performative love for whales relates to our relentless extermination of them'.

Whales also continue to be harmed by man-made noise. It began to become a serious matter with the din of propeller-driven ships in the nineteenth century and has been ratcheting up ever since as ships have grown in size, number and speed. Increased military activity, the use of sonar, seismic airgun surveys for oil and gas, and offshore construction have added to the din. There has been a nearly fifty-fold increase in global container ship capacity since 1974, and it is estimated that noise levels in the oceans have doubled every decade since the mid twentieth century. Swimming under or near a passing super-tanker is like being under the path of a large jetliner as it takes off. As a consequence of all this, it has become harder for many baleen whales to be heard when they sing. The communication range of a blue whale is thought to have shrunk about ten-fold since the 1940s. Toothed whales such as sperm whales and orcas, which hunt using echolocation, have also suffered. In the noisiest places their hearing range has reduced by as much as 95 per cent. 'We are injecting so much noise that we are acoustically bleaching the ocean,' says Clark in the 2016 film *Sonic Sea*. 'All the singing voices of the planet are lost in that cloud of noise.' Many of the recordings of the uninterrupted whale song that so many people have come to love were made when the oceans were orders of magnitude quieter than they are now, and because of this they are not representative of the world that many whales actually experience most of the time.

The good news is that sound pollution in the sea can be reduced. Seismic airguns can be stood down when whales and dolphins are migrating. Changes in design and lower speeds can reduce ship noise. When the volume of shipping fell sharply in the first months of the pandemic in 2020, parts of the ocean adjacent to sea lanes became markedly quieter. The extent to which this lull benefited whales and other marine life is unclear, but there are grounds for hope. Earlier studies have shown that noise reduction in the days after 11 September 2001 resulted in much lower stress in North Atlantic right whales. Quite small shifts in the sea lanes plied by oil tankers and container ships near Sri Lanka could reduce both noise and ship strikes that affect blue whales in those waters. (See the entry in this book on 'Frontiers' for a hopeful story.)

There is more to discover, for ill and good. It was reported in 2020 that Chinese scientists were studying how to disguise underwater military communications as dolphin and orca clicks and killer whale songs. Elsewhere, researchers have found that the songs of fin whales can penetrate more than two thousand metres down into rock beneath the ocean floor. Listening in, they can use the seismic waves that these songs create to study the underlying structure of the planet. Meanwhile there are new riddles. The songs of blue whales, for instance, have become much deeper over the last forty years, declining by around 30 per cent in pitch, and no explanation suggested so far – whether climate change-related fluctuations in ocean acidity, shifts in average whale size and population density, or rising ocean noise – seems to fully account for the change.

'Whales hold a history in their song that we don't know about,' Roger Payne said in a radio programme first broadcast in 2020. 'Isn't there some different way that humanity could interact with the wild world than by just desecrating it?' If there is, it will require acts of collective human imagination that include a deeper sense of the past and future of life on Earth. Only humans, suggests Rebecca Giggs, 'have concepts of the past occupied by whales and their ancestors. We are the animals able to envision the time to come and the nature that will abide it.'

Perhaps. In any case it may not be too late to learn from the Māori tradition of Aotearoa/New Zealand, in which the whale – the largest sea creature – is the 'oceanic twin' of the *kāuri* – the largest native tree. Both are regarded as *rangatira* (chiefs), as respected *tupuna* (ancestors) of Māori, and as *taonga* (treasured, sacred) species. 'Song is life,' says Clark. 'It is the essence of who we are, and it joins us all.'

Leviathan, or the Sperm Whale

'Is it not curious,' asks Herman Melville in *Moby Dick*, 'that so vast a being as the [sperm] whale should hear the thunder through an ear which is smaller than a hare's?' But Melville was wrong. For all their expertise in butchery, the whalers of his time had an incomplete understanding of sperm whale anatomy and, noting the tiny external ear on the side of its head, overlooked what was actually right in front of them: one of the largest and most astonishing organs of perception that has so far evolved on Earth.

The long bulbous 'head' of a sperm whale is actually a giant hooter: a nose that has evolved to become an ear, a range finder and an organ of speech. Melville asks whether the whale would be longer of sight or sharper of hearing if its eyes were as broad as Herschel's great telescope or its ears as capacious as the porches of cathedrals? And, contrary to his assumptions, the answer is yes. This huge nose, half the size of a bus – and so several times larger than the 120-centimetre (forty-eight-inch) mirror of Herschel's optical tube, and so the equal of some cathedral doors – enables the sperm whale to project sounds and detect echoes with precision and at distances unmatched by any other living being apart from humans equipped with high technology. It is, says marine biologist Hal Whitehead, 'the most powerful sonar in the natural world'.

In the kind of spectacular kludge that only evolution could dream up, the whale's left nostril has become its single blowhole sitting at the end of a tube several metres long at the very front and slightly off-centre of the top of its massive head. When the animal dives it closes the blowhole and instead drives air through a pair of sonic 'lips' situated just beneath and inside. Clicking sounds made by the lips are projected back all the way through a huge horizontal organ unique to sperm whales that is filled with thousands of litres of spermaceti, a fine oil with superb acoustic properties, before bouncing off a sound mirror called the frontal sac, which is mounted on a gently sloping dish of bone just above the whale's eye and brain case right at the back of this extraordinary nose. The reflected clicks are then focused forwards again through a lower horizontal compartment filled with what is misleadingly known as 'junk' and projected out into the sea. The whale can vary the power and rapidity of the clicks at will. Some click sequences are for echo-location: vibrations that bounce off prey, seamounts and other objects, and are then picked up again in a lozenge-shaped acoustic fat pad at the back of the whale's jaw and passed on to its ears. Other clicks are for communicating with other sperm whales.

Sperm whales are the largest toothed whales, a group that includes dolphins, orcas, belugas, narwhals and beaked whales. Females typically grow to around eleven metres (thirty-six feet), or about the length of an old-fashioned double-decker bus, and males to sixteen metres (fifty-two feet). Together with their pygmy and dwarf sperm cousins, they are the only surviving members of what was once a large family. One of their extinct cousins, whose remains were discovered in a desert in Peru in 2010, was much the same size and shape as a modern sperm whale but had a far thicker and stronger lower jaw, lined with enormous teeth. Its finders named it *Livyatan melvillei* in a fusion of the name of a fearsome primordial sea monster with that of the author of *Moby Dick*. Like the giant shark *Megalodon*, with which it shared the Southern Ocean around 9 million years ago, *Livyatan* fed on small baleen whales.

Modern sperm whales also need prodigious quantities of food, but for them the fare is mostly squid. Every day, a typical adult will put away 700 to 800 of these cephalopod snacks, most of them about a foot long, as well as much larger ones when it can. To adapt Monty Python on spam: squid, squid, squid, squid, squid, squid, lovely squid, beautiful squid. But also, deep, dark squid. To satisfy their hunger sperm whales have to dive hundreds and even thousands of metres down to search through a huge, cold expanse in almost complete darkness. The deepest dive logged so far is over 1,200 metres, or about 3,900 feet, and researchers think they may descend to twice that depth. Adults spend about half their lives down here, continuously foraging by emitting steady clicks at intervals of about half to one second. The sound resembles a powerful, insistent metronome: chock, chock, chock.

When a whale identifies a target – a squad of squid – the clicks come faster and faster until they are so close together that the human ear, listening through a hydrophone, hears them as a creaking sound, like a rusty door. This close spacing in 'click trains' of 600 times per second or more enables the animal to build up a more detailed picture-in-sound of what is in front of it. Smart tags placed by marine biologists on diving whales to record sounds and movements show the whales accelerating, twisting and manoeuvring underwater. When clicks are followed by silence rather than a return to the normal slow click, the whale is (it is presumed), raking up squid with its long lower jaw and gulping them down. Yum.

The blue whale may be the largest of all the animals, but sperm whales are the loudest. The clicks they produce when hunting can be at well over 200 underwater decibels. This is more than enough to burst the eardrums of a human diver at close range, and could even vibrate them to death, and it has been suggested that sperm whales might use their loudest clicks to stun or kill prey. The evidence is, however, that in the 'terminal buzz phase', when the intervals between clicks are very short and the whale is homing in on its lunch, the clicks are actually more than an order of magnitude below the levels that would debilitate or kill. Loud clicks are more likely to be used for long-distance scanning, enabling a sperm whale to detect squid from hundreds of metres away, or to signal their presence to other whales.

It may be that the loudest animal sound in the ocean has driven the evolution of the one with the largest and most sensitive eyes. For sperm whales do not only prey on small fry. Giant squid, which can grow to well over ten metres in length, are a dainty dish if you can catch them, but with eyes bigger than a human head, these colossi can probably scan the dark for the faint flashes of bioluminescence created by a sperm whale as it speeds through the depths and bumps against small jellyfish, crustaceans and other plankton.

A typical sperm whale dive lasts about forty minutes, although it can be over an hour, and is interspersed with ten- to twelve-minute intervals to breathe. Meanwhile the babies, who have not learned to dive, stay near the surface in the company of either their mother or another adult female, who keeps an eye out for predatory orcas or other dangers while the mother feeds in the deep. (Sperm whales have excellent hearing so babies and young ones can probably follow what is going on with the hunt far below.) Once a day or so, all the whales in their group, which is known as a pod, will gather together on the surface, sometimes for hours on end, and in these circumstances they use their clicks for a very different purpose.

Sperm whales are intensely social animals, and like to catch up and reinforce their bonds whenever they can. They have the largest brains on Earth – five or six times larger than those of humans – with two key markers of a capacity for complex thought and feeling: abundant spindle cells, and a very well developed neocortex. The writer Philip Hoare describes an encounter on a dive in the Azores: 'six or seven animals . . . spent hours rolling around one another in what biologists call a socially active group. Their ebony-black or dove-grey bodies were in contact throughout. Fins stroked flanks, gentle jaws bit one another. At one point, two of the whales brought their pugnacious, square foreheads together in what looked like an act of philosophical communion.' They were, he writes, 'communicating with one another in the most intense, sensual manner'.

As they engage in this physical contact, the whales generate patterns of clicks that researchers call codas. Unlike the clicks made while hunting, these are organised rather like Morse code, with three to forty or so clicks interspersed with pauses. Codas are signals of identity, and ways to share information. Hal Whitehead, who starting diving with sperm whales at a time when many people feared it might be dangerous to do so, once heard a particularly heavy burst of codas as a new baby was being born. Like humans and other apes, as well as dolphins and some birds, sperm whale babies babble at first, and it takes them a couple of years to learn the distinctive codas of their pod. Certain codas are shared not just with family and pod but with members of a larger and more widespread clan, which can comprise thousands of individuals. There are at least five distinct clans across the Pacific, and they differ according to their codas, just as dialects or languages differ between groups of humans. Sperm whales seldom if ever mix with members of other clans, but they do seem to communicate with distant members of their own clan, and this kind of information-sharing may explain how so many of them learned quickly how to outmanoeuvre hunters in sailing ships in the nineteenth century, and in recent decades to share techniques for stealing fish from hooks on long lines dragged behind boats.

Sperm whales live in what researchers call a 'mother culture', in which a group of adult females led by a matriarch take care of the young of both sexes. A good terrestrial parallel is elephants. Just think of giant, super-sentient, deep-diving elephants. 'The main thing I've learned from [sperm] whales,' the researcher Shane Gero tells the author and biologist Carl Safina, 'is that your family is the most important thing. Learn from Grandma; love your mom; spend time with your siblings.' Spending so much time with the whales has changed Gero's world. 'Learning what the whales value has helped me learn what I value.'

While female and young whales stay in the warm waters of the tropics, adult males spend most of the year foraging at higher latitudes. But the males are by no means excluded from the pod, and when they return they announce their arrival with extremely loud and slow clicks or 'clangs'. To humans, this noise can seem eerie. Whitehead suggests it could be behind the story of Davy Jones, the ghostly seaman who knocks to be let out of his watery grave. But for female sperm whales these noises are, apparently, extremely attractive, and when they meet the males are 'all calm serenity and gentleness'.

Free divers who slip into the water without scuba or other noisy gear and have been able to swim close to whales talk of the awe and connection they feel in the unmediated presence of such intelligent and fully conscious beings. In one such dive, Philip Hoare describes a large female swimming directly up to his comparatively tiny body and pausing to examine him, her sonar reverberating – click-click-click – through his skull, sternum and whole skeleton. 'It was ironic. As a writer, I'd spent years trying to describe whales, and here was a whale, trying to describe me . . . She looked at me with absolute sentience and curiosity, wondering what I was.'

The world we make depends on what we choose to do. Although humans have almost entirely stopped hunting and killing sperm whales, they are still classed as being at risk of extinction. Entanglements with fishing gear, high levels of man-made contaminants in their bodies, and the historic loss of knowledgeable older adults all take a toll. Noise pollution from shipping and seismic surveys disorients and traumatises them. On the plus side, there may be several hundred thousand still alive, mostly far from land and thriving on food sources that are not yet all dangerously contaminated or depleted by humans. It is possible that these animals will continue to enrich the world and its soundscape for millennia to come – and even that humans will begin to understand a little better what they are saying to each other, and what they can teach us.

Blackbird

Epiphanies are few and far between when you're putting out the bins on a cold night in February. But beyond the scraggly bush and the streetlight, a blackbird was pouring out notes so lovely that I stopped in my tracks. A gorgeous flute-like sound, and very loud; a free jazz flow that never seemed to end. I found it impossible not to believe that the bird was feeling intense joy. 'Where does it come from, the song?' wondered D. H. Lawrence in an essay written in 1917. 'After so long a cruelty [of winter], how can they make it up so quickly?'

In *Birds Britannica*, a compendium of natural history and folklore, Mark Cocker and Richard Mabey report that many people in Britain find the song of the blackbird to be equal if not superior to that of its cousin the song thrush. The many attempts to convey its quality have, they write, thrown up a consistent pattern of description: 'a certain languid and easy delivery'; 'lazy and gives the impression of sleepy contentment'; 'trilled out . . . as if by being at peace and supremely happy'.

The blackbird's song is made possible by an anatomical marvel. In humans, the larynx, or voice box, sits high in the throat at the top of the windpipe, but the equivalent organ in birds is at the bottom of the windpipe, just above the lungs. Known as the syrinx, it is named after a nymph who fled from the god Pan and was transformed into a reed, and thence a flute. It actually consists of a pair of identical organs within the bronchi, the twin tubes that connect the windpipe to the lungs. Inside each tube a circular, flexible membrane called a tympanum faces a small bump of erectile tissue on the opposite wall of the tube. The bird adjusts the diameter of the tympanum and how far the tissue protrudes into the tube as it forces air through, creating changes in pitch and timbre that become parts of its song. The double nature of the syrinx enables some species of birds to sing two different notes at the same time.

In addition, the respiratory system of birds allows them to sing for a long time almost without tiring. Unlike humans and other mammals, who simply breathe in and out, birds have large air sacs inside their bodies that enable them to pass oxygenated air through their lungs continuously in one direction. The sacs act like bellows, first drawing air rich in oxygen into a set at the rear of the body before passing it forwards through the lungs and then into another set at the front of the chest, whence it is expelled on the next out-breath. This arrangement, combined with the fact that the syrinx can produce sound on both the in- and out-breath, is what makes it possible for a hovering skylark to pour forth its full heart in profuse strains, as well as my bin-based blackbird beatitude.

Studies of blackbirds in the Białowieża Forest in Poland, which is one of Europe's oldest woodlands, suggest that they were originally creatures of the high treetops. Their dark feathers and low-frequency songs, which travel quite well through thick vegetation, are very like those of other members of the thrush family, who thrive in the canopy of tropical forests. There is, perhaps, an echo of this history of long-distance calling in Edward Thomas's 1914 poem 'Adlestrop', in which a blackbird calls to others far across the countryside of Oxfordshire and Gloucestershire.

The number of many kinds of birds in Europe, the United

States and elsewhere is falling fast. There are hundreds of millions fewer than forty years ago. But blackbirds seem to be among those that are most adaptable and resilient, at least for now. Even as many rural areas drain of wildlife, they are thriving in modern urban environments. The majority of the 5 to 6 million pairs in Britain today live in the cities and suburbs, where they are found at up to ten times the density of farmlands that are often laden with toxic chemicals and depleted of the invertebrates and other life on which the birds feed. Their determination is impressive. It was probably a city bird that inspired Paul McCartney's song 'Blackbird', written in support of the Civil Rights Movement in the United States shortly after the assassination of Martin Luther King Jr in the spring of 1968. More recently, Cocker and Mabey report how, over one Easter holiday, a blackbird nested in the engine compartment of a forklift truck on an industrial site in Colchester, Essex. 'When we returned on Tuesday,' says their local correspondent Alex Dunn, '[she] continued to sit as work resumed, being driven around the site from 7.30 a.m. until 4.30 p.m. on a daily basis. Five eggs were laid and four chicks were hatched, which the hen fed by leaving the vehicle whenever it stopped going forwards, and returning with food when it was stationary.' In 'St Kevin and the Blackbird', Seamus Heaney describes the holy man holding out his hand in the sun and rain for weeks on end to hold the bird's eggs safely until the young are hatched, fledged and flown; but sometimes if you're a blackbird and there are no saints to hand you have to make do with a forklift truck.

Whether or not one is stumbling around in the dark with the bins, the song of a blackbird can remind a human listener that something larger than the self endures. Writing in his diary from the Western Front in early March 1917, three years after 'Adlestrop' and a month before his death at the Battle of Arras, Edward Thomas noted the 'chinking' of blackbirds during a lull in the bombardment. That same spring D. H. Lawrence was struck by how song 'bubbles through' the birds: 'they are like ... little fountain-heads whence the spring trickles and bubbles forth ... In their throats the new life distils itself into sound.' And Lawrence's ecstatic intuition may capture a deep truth. Some ecologists argue for an understanding of the living world as an ensemble in which – in an apparent inversion of common sense – there is not a song because there are singers, but singers because there is a song. The idea is that large-scale processes and living systems favour the evolution, or recruitment, of species that are capable of thriving by fulfilling roles in that larger process. Indeed, it may be that, as the biochemist Nick Lane puts it, 'it is the movement that creates the form' for life itself. 'I am the song that sings the bird,' wrote the poet Charles Causley. Life calls to us even as we call to it.

Owl

Glory be to Creation for wonky things – for the outsize one-claw of the fiddler crab, the narwhal's left-lip tusk-tooth and the side-skewed bent beak of the wrybill. Land and sea are charged with things counter, original, and strange. Mid-ocean, the strawberry squid, its pink mantle dotted as if with seeds, turns one big eye upwards to scan for predators or prey outlined against the light in the water above, while a smaller eye looks down for pinpoints of glowing life. At the sea bottom, from coastal shallows down to the deepest trench, hundreds of varieties of flatfish – the Pleuronectiformes – stare up with a second eye that has yanked itself all the way round their skulls.

I must admit a bias here, for I am among the wonky creatures going to and fro on the earth and walking up and down upon it. My outer ears, or pinnae, are lopsided – a relatively common (and harmless) condition, but one that becomes more noticeable with age, and sometimes I wonder if I'll wake up one morning to find that, like a baby flatfish, my other features are starting to morph and migrate around my skull as well.

The asymmetry of my outer ears is external and accidental, and it does not affect how I hear, but there are creatures for whom wonkiness is a feature of their hearing system, not a bug. In most toothed whales, for instance, the skull is squashed to the left side to make room for soft tissues that help them echolocate. And in some species of owl, the openings of the ear canals (which are located not in the tufts that some species have on top of their heads but on either side of the skull) emerge at slightly different heights. This difference in position means that sounds from above and below reach the left and the right ear at slightly different times and are fractionally louder or softer, enabling the owl to better identify their origin in the vertical as well as the horizontal plane. To enhance this, the feathers on an owl's face are shaped rather like a parabolic reflector and help focus sound on to each of the differently placed ears. It is as if the birds have a large ear trumpet mounted on the front of the head, which they can move at will to better focus sound. The downwards-facing, sharply triangular beak also minimises sound reflection away from the face. Further, an owl's hearing does not decline with age because, like many other birds but unlike humans, the sensitive hair cells of the inner ear regenerate continuously. Go, owls!

There are about 300 kinds of owl, and they are a diverse lot. Blakiston's fish owl, the largest, has a wingspan of over two metres, and thrives (or did) in the snowy north of Japan as well as parts of the north-east Asian mainland that are also home to tigers and bears. The elf owl, which is the smallest, is the size of a sparrow and weighs just a hundredth as much. It lives in the drylands of south-west North America, where it nests in holes in giant cactuses that have been carved by woodpeckers.

Owls have been a significant presence in the human imagination for a long time. The engraving of a long-eared owl in the Chauvet Cave in France is about 32,000 years old. It is shown with its head seen from the front but its body from the back – a posture that is easy for owls, thanks to the fourteen vertebrae in their necks compared to our seven. We can only guess what the image meant to those who made it, but it is clear that owls have symbolised different things for different cultures at different times. The Wardaman people of Australia believe that Gordol the owl created the world, while the Nyungar protect an 'owl stone': a standing rock that represents Boyay Gogomat, the great creator, healer and destroyer. The Maya, Aztecs and other peoples of Mesoamerica considered the owl to be a symbol of death, while in European folklore they were long thought to be birds of ill omen. (Owls were widely believed to be familiars of witches, and it was said they had two nipples like a human mother and would suckle a newborn baby at night; garlic was sometimes placed on a sleeping baby to ward them off.) By contrast, the Navajo say that owls and coyotes hold the balance of day and night, while for the Sami of northern Scandinavia owls bring good luck, and the Ainu of northern Japan revere Blakiston's fish owl as a guardian spirit. In ancient Greece the owl was a symbol of Athena, the goddess of wisdom, while in India an owl (along with an elephant) is the mount of the goddess Lakshmi, who is considered variously as the goddess of fortune, power, beauty, the pursuit of ethical life, understanding and liberation.

A major factor in the resonance of owls in our psyches, surely, is their large, forwards-looking eyes, which are suggestive of vigilance and attention. The poet Eliot Weinberger reports multiple sightings in medieval England of birdlike 'angels', often 'of a bluish and about the bigness of a capon, having faces like owls'. But no less wondrous is the almost complete silence of their flight. Inaudible to humans until they are within little more than an arm's length, owls arrive and depart like ghosts or spirits.

This sonic marvel is thanks in part to an owl's body plan. Even the gentlest flap of a bird's wing makes some noise, but because owls have especially large wings relative to the size of their bodies they can glide further and flap their wings less often than other raptors such as eagles and falcons. Larger wings and fewer flaps also mean that owls can fly very slowly, which reduces air turbulence and the noise arising from it. Barn owls, for example, can fly more slowly than humans walk. This allows them more time to listen out for prey such as small rodents, and locate them with precision in darkness or beneath snow or vegetation. Thanks to this ability barn owls thrive in a wide range of environments on every continent except Antarctica.

The remarkable shapes of owl feathers also help in at least three ways. First, comb-like serrations on the leading edge of the wing feathers break up turbulent air that would otherwise create a swooshing noise. Second, these smaller streams of air are further dampened by a velvety texture that is almost unique to owl feathers. Third, a soft, ragged fringe on feathers on the wing's trailing edge helps to diffuse turbulence behind the wing. And all this is almost certainly not the whole story. It may be, for example, that the velvety texture of owl feathers and their shaggy fringes reduces frictional noise between wing feathers when they are moving.

Some of the first systematic research into how owl feathers reduce noise was undertaken in 1934 in the hope that it could inform the design of quieter propellers for aircraft. Little was achieved at the time, but researchers today aim to learn more. The results will not necessarily all be benign. Silent drones could be more deadly weapons, but on the positive side new materials inspired by owl feathers may one day reduce noise and energy use, with significant potential benefit to human and animal well-being. If so, it would be a small example of biomimicry, or design modelled on living processes, which can radically increase efficiency and reduce waste as part of a systems approach.

In the 1982 film *Blade Runner*, an animoid, or artificial, owl (which is played on screen by a real, living Eurasian eagle owl) looks on indifferently as the replicant Roy Batty gouges out the eyes of his Frankenstein-like creator Eldon Tyrell. Had

Tyrell taken a different path, and not treated his creation as merely a tool to work to death for profit, he might have escaped such a fate. But what lesson, if any, is hidden in the image of the owl engraved in the Chauvet Cave? Like the Angel of History imagined by Walter Benjamin, the Chauvet owl flies into the unknown with its eyes and ears facing backwards. So we humans hurtle into a future we cannot see or hear, but which we shape.

Nightingale

It is hard to improve on a description of the song of the nightingale by the author H. E. Bates. It has, he wrote, 'some kind of electric, suspended quality that has a far deeper beauty than its sweetness,' and is 'a performance made up, very often, more of silence than of utterance. The very silences have a kind of passion in them, a sense of breathlessness and restraint, of restraint about to be magically broken.' A nightingale's song, said Bates, 'can be curiously seductive and maddening, the song beginning very often by a sudden low chucking, a kind of plucking of strings, a sort of tuning up, then flaring out in a moment into a crescendo of fire and honey and then, abruptly, cut off again in the very middle of the phrase. And then comes that long, suspended wait for the phrase to be taken up again, the breathless hushed interval that is so beautiful.'

This is right about so many things, not least the silences. I particularly like 'a crescendo of fire and honey'. But it's striking, too, that Bates, who was writing in 1936, does not make analogies that are frequent today, such as the similarity of the scratches, pops and warbles in nightingale song to electronic music. And to my mind he misses, or at least understates, three things that have struck me on the occasions that I have heard a nightingale myself. The first is just how loud they can be. In an effort to be heard over the top of city noise nightingales can sing at up to ninety-five decibels – as loud as a chainsaw or a large truck, and loud enough to break European noise pollution regulations. The second is just how complex the song is, drawing on a repertoire of over 200 distinct parts, each lasting from three to twenty seconds, which the bird combines in precise and different ways that as yet elude human understanding. (The Finnish name for the bird is particularly apt: *satakieli* means 'hundred voices'.) The third is just how strange they can sound. To adapt what Louis MacNeice wrote in his 1935 poem 'Snow', the song of the nightingale is crazier and more of it than we think, incorrigibly plural.

None of this is to say that the responses of earlier generations to the nightingale are irrelevant. If anything, the opposite is true. It is wondrous to glimpse the yearning that Sappho felt twenty-six centuries ago in Greece when she wrote of the nightingale as the 'herald of spring / with a voice of longing', and the bliss of Hafez in fourteenth-century Persia when he exclaimed, 'the nightingales are drunk!', and described a complete happiness that welcomes all, as well as a recognition that there can be no happiness without sorrow. John Keats's 1819 poem 'Ode to a Nightingale' is among the most powerful statements in English of what it is to apprehend surpassing and enduring beauty in a life that is far too short.

At the same time, these and other accounts, including Bates's, remind us that some things are specific to time and place, and that as these change so does what and how people hear. Consider attitudes in Britain. The number of breeding pairs of nightingales in these islands, which are on the north-western edge of a range that extends to Mongolia in the east, has declined steeply in recent decades, and the species has been placed on the red list for conservation. Increasingly rare, nightingales have become an icon (an earcon?) for natural beauty threatened with extinction. Recognising this, in April

2019 members of Extinction Rebellion turned 'A Nightingale Sang In Berkeley Square' into an unlikely protest song. At a chill-down held in Berkeley Square after the high-intensity blockades of the previous weeks, hundreds of activists sang the song, a schmaltzy number written in 1939 and associated with Vera Lynn and the Blitz, while the street artist ATM displayed a painting of a nightingale on the tailplane of a Lancaster bomber.

Behind the performance was history as myth and model. On 19 May 1924, the cellist Beatrice Harrison, who had given the first festival performance of Elgar's tragic Cello Concerto a few years earlier, made one of the BBC's earliest live outside broadcasts from her Surrey garden, duetting with a nightingale. Or so went the story until it was revealed in 2022 that on the day itself BBC engineers galumphing about with heavy equipment had spooked the bird away, and a variety performer – a whistler or *siffleur* known as Madame Saberon, but whose real name was Maude Gould – was called in at the last minute to mimic its song.

All the same, the combination of cellist and 'bird' was a hit, and the broadcast was repeated in subsequent years. When, however, the BBC came back in 1942 to broadcast the nightingales without Harrison, their microphones inadvertently picked up the thrum of 200 Wellington and Lancaster bombers passing overhead on their way to raids in Germany. An engineer realised the sound might warn Nazi air defences, and pulled the plug. In calling on these memories the activists of 2019 were summoning a structure of feeling and a network of beliefs embedded in the national imaginary: a world in peril, and the need for courage and determination. The folk singer Sam Lee and the Nest Collective, who convened the Berkeley Square event, continued to combine music and environmental concerns in an ongoing series of encounters, *Singing with Nightingales*, in the Sussex woods.

Across much of continental Europe and beyond, however,

where nightingales are thriving, the resonances are different from those in Britain. In Berlin, the males that have recently arrived from West Africa on their spring migration sing in parks in the city centre and even at noisy traffic intersections. The explanation is simple: the city retains large areas of relatively untended, 'messy' vegetation which resemble scrubland – successional habitat in a state of dynamic transition from open grass or heath to wood that is favoured by nightingales because it provides hiding places from predators. At any rate, the fact that there are lots of nightingales means that people in Berlin hear them more often than people in Britain, and the city has become an excellent place to study and appreciate their songs.

Charles Darwin suggested that when male birds sing, females appreciate certain characteristics of their songs, such as their complexity, because, just like humans, they have a sense of beauty. The idea has not been disproven – although ornithologists have recently discovered that females of many species of songbird sing too. It is only in places such as northern Europe that migrating males, arriving first, are the ones to sing as they claim territory. But Darwin's hypothesis begs a question: *why do females find these characteristics beautiful?* And in any case, beauty is unlikely to be the whole story: complex or loud song may be a sign of intelligence or strength. Studies of nightingales in Berlin have found a correlation between song complexity in males and the extent to which those males later help females feed chicks. 'The song,' suggests Conny Landgraf of the Leibniz Institute for Zoo and Wildlife Research, 'is like a promise that the males give to the females to be good fathers.'

This does not mean that female nightingales are unemotional and merely seeking to maximise parental utility. 'It is clear that nightingale song [conveys] emotions,' says Dietmar Todt, who has spent a lifetime studying animal communication, and it would be surprising if songs didn't have an effect on both the singer and the listening birds. Indeed, it seems likely that the impact is more powerful for the birds than it is for humans given how much louder their songs are in relation to their size and how much more intricate. I know from my own experience how a good sing, and hearing other humans sing, can make me feel, and can only imagine how this scales up if you're a nightingale.

Is nightingale song music? Tina Roeske of the wonderfully named Max Planck Institute of Empirical Aesthetics thinks so: 'they perform for each other!'. Both she and Todt appear in a film with the musician and philosopher David Rothenberg that focuses on his forays into making music with the birds. *Nightingales in Berlin* features Rothenberg on clarinet and fellow musicians – including a singer, a violinist, and an oud-player and others on thumb piano, bodhran and synthesiser – listening intently to the birds in Berlin parks and responding with free jazz improvisations. 'It's so strange . . . You attend to the [nightingale's] song, and at the end when you get really into it, you feel like you're flying,' says the singer Lembe Lokk after her duet.

For Rothenberg, 'anyone who studies music should study more than human music [and] investigate how human music builds on the natural music that's been there for millions of years'. And in this he is part of a long line. Pipe- and flute-players have surely been listening to birdsong for millennia. In the Western classical tradition, an attempt to notate birdsong by Athanasius Kircher in his *Musurgia Universalis* of 1650 inspired Heinrich Biber to write a *Sonata Representativa* in 1669 that includes an imitation of a nightingale on the violin. In *Le rappel des oiseaux*, a 1724 work for keyboard, Jean-Philippe Rameau creates a plausible, though generic, birdsong-like effect in the upper voice. Many composers and musicians have followed. Maurice Ravel's *Oiseaux tristes*, composed in 1904–5, harks back to Rameau but explores new harmonic worlds. Olivier Messiaen turned to birdsong throughout his life, notably with *Abîme des oiseaux* for solo clarinet in the *Quartet for the End of Time*, and in the *Catalogue d'oiseaux*, a seven-book set of solo piano pieces. Of the former, Messiaen famously said, 'the abyss is time with its sadness, its weariness. The birds are the opposite to time; they are our desire for light, for stars, for rainbows, and for jubilant songs.' Musicians today, including Rothenberg, often seem to be motivated by a similar desire to access a different relationship with existence through engagement with birdsong, which is not actually timeless but is vastly more ancient and also – in the case of the nightingale and some other birds – moves faster than humans easily comprehend. In *Bird Concerto with Pianosong* (2001), Jonathan Harvey 'stretches' birdsong into the human realm, slowing it down so that a pianist and other musicians may interact with it. Nightingales are not among the forty or so species, but there is surprise and endless creativity throughout the work.

Beautiful as human music inspired by birds can be, there is always a gap between them and us, because songbirds hear and process sound differently from humans. While birds generally hear the same range of pitches as us, many also hear details that are imperceptibly fast for humans: they are capable of registering units of sound that last as little as one millisecond, while humans can manage three to four milliseconds at best. Further, it seems that many songbirds don't care so much about the sequence of notes as their individual quality. This has been compared to humans paying close attention to the nuance of each other's vowels while ignoring the order of each other's words – an almost unimaginable thing. Birds such as nightingales 'appear to listen most closely not to the melodies that catch our ears but rather to fine acoustic details in the chips and twangs of their songs that lie beyond the range of human perception,' writes the cognitive scientist Adam Fishbein. Outside of works for the shakuhachi and genres such as *musique concrète*, little human music follows this route, and when it does it has to take account of our much slower ears.

According to David Chalmers, a philosopher who is probably best known for coining the phrase 'the hard problem' to describe the challenge of explaining consciousness, advances in technology will deliver virtual worlds that rival and then surpass the physical realm. With limitless, life-like experiences on tap, he suggests, the actual material world may lose its allure. A similar point has been made more provocatively (or crudely) by the tech investor Marc Andreessen: 'reality has had five thousand years to get good, and is clearly still woefully lacking for most people.' For some, the future is the metaverse – in the vision of Mark Zuckerberg, the CEO of Meta (formerly Facebook), a seamless space where fish swim in the trees and your favourite friends are always within reach. But for its critics the metaverse as currently conceived is full of dangers. Think 'virtual reality with unskippable ads,' says Wendy Liu, a writer on tech. 'For its makers,' quips the journalist Anna Wiener, 'the metaverse will be stuffed with money – in every dimension, all the way down.'

Zuckerberg's vision is easily mocked (search online for a skit about the Icelandverse, for example), but there are serious questions here. Andreessen's dismissal of the natural world, whether or not he was being entirely serious, is in line with a colonial mentality that has turned much of the world into a standing reserve of resources that can be extracted without consequence. 'Only once we imagined the world as dead could we dedicate ourselves to making it so,' writes the author Ben Ehrenreich of this mentality. And whoever controls the platforms – whether corporations, governments or other actors – there is an underlying mismatch that bodes ill. For the foreseeable future, tech will be software running on hardware, and this is quite unlike life itself, which *is* a process of self-organisation and cognition (though not, I think, consciousness) all the way down.

Against the supposed inevitability of the metaverse, we

might hazard what the novelist Amitav Ghosh calls a 'politics of aliveness'. And I'd like to think of the song of the nightingale as part of the soundtrack for this. It is a fractured music; a song sung in darkness, but a darkness of possibility, not death. Endlessly evolving and beyond human control or understanding, it is, as Rothenberg puts it, 'always more and less than anything we can add to or take away from it'.

I return in memory to the rare occasions when I have heard a nightingale for myself in England. Once was at Fingringhoe Wick, a former quarry that is now a nature reserve: an untidy but beautiful place that writers such as Richard Mabey and Stephen Moss would call unofficial or accidental countryside. Another was at Knepp, a farm where formerly manicured pasture is turning into scrappy, fertile 'mess' as part of a rewilding project that has resulted in an astonishing bounce back in non-human life. On both occasions I was struck more by the sheer volume, energy and strangeness of the song than anything I experienced as beauty. And I reflect on the fact that the abundance of nightingales, for now, in a big city like Berlin is a reminder of a victory (pyrrhic or otherwise) over a death cult that would have made the German capital the centre of a monstrous plan for world domination. Instead, Berlin today is a dynamic, slightly ramshackle place, full of people from everywhere, fizzing with ideas and creativity, as well as dangers.

In the poem 'Night Singing', W. S. Merwin asks what can be said about the song of the nightingale after so many poems across the centuries and so much scientific study? He finds he can 'only listen' and ride out on the 'invisible beam' of the bird's long note that 'wells up and bursts from its unknown star'. Meanwhile, he writes, actual starlight glitters among the small leaves of May. This meeting of the song of the bird with the light from other worlds as apprehended by Merwin seems to me just right. Nightingale song has evolved over millions of years, and is as old as starlight. 'The planet will never come alive for you,' writes Ghosh, 'unless your songs and stories give life to all the beings, seen and unseen, that inhabit a living Earth.'

ANTHROPOPHONY
Sounds of Humanity

Rhythm (3) – Music and Dance

One of my daughter's favourite games when she was very small was to bounce on my knee in time to a nursery rhyme that begins 'This is the way the ladies ride: bumpety, bumpety, bump.' With each different character who was riding (the gentleman, the farmer), the speed and manner of the bounce would change, until in the final line I would say her name, pause, and then . . . bounce her at crazy and chaotic speed – '*bumpety bumpety bumpety bumpety bumpety bumpety!*'. Then we would do it all over again about ten times.

A vanilla definition of rhythm is that it is any regular, repeated pattern of sound or movement. It involves, 'a being-taken-hold-of . . . that has always already occurred,' writes Vincent Barletta. For sure, it starts early. A foetus in the womb begins to sense its mother's heartbeat within about eighteen weeks of conception, and newborns recognise and show a preference for regular beats from their first day of life, and will move along with them – although it can take until they are about four years old to stay consistently on beat. Games like the one I played with my daughter, and which millions of parents and children play every day, help us to learn how rhythm works, and how we can join, or 'entrain' with, others by moving or singing in time – or varying our response. They help teach about anticipation, turn-taking, cooperation and playful surprise. And we are able to keep learning and expanding these capabilities throughout our lifetimes. A human being, says the dancer Kimerer LaMothe, is 'a rhythm of bodily becoming'. Maybe Friedrich Nietzsche put it best: 'I would only believe in a god who knew how to dance.'

Some animals use rhythms in sound to communicate with others. When a male bush cricket wants to attract females he joins a large group and, in the audible equivalent of the synchronous flashing of fireflies, chirps at exactly the same time as his fellows. The simultaneity creates amplitude summation: together the crickets are louder and can attract females from further away. Once the females are close, however, individual males start to compete. Using the common pulse as his baseline, a male cricket will try to signal a fraction of a second ahead of the others in a bid to stand out. It's been described as similar to the way jazz musicians swing and syncopate their beat – but with the difference that the cricket's behaviour is fixed: it never learns a new beat. This phenomenon may be very ancient. Fossils show that a katydid in the same family as bush crickets alive today was stridulating to a regular beat 165 million years ago.

There are also distinct rhythmical patterns in many bird calls and songs. The writer and musician Rhodri Marsden has recorded a collared dove on the chimney of his house cooing in 5/4 time and at exactly the same speed as 'Take Five' by The Dave Brubeck Quartet. But, remarkable as they are, many bird calls and songs change little, and although some birds do identify rhythms in the sounds of others and alter their own in response to some extent, evidence suggests they are mainly listening to timbre and microtone. Songbirds such as starlings and lyrebirds are superb mimics of other species, as well as of random human sounds such as phones and car alarms (and, in the case of lyrebirds, uncannily accurate imitations of crying human babies), but they do not sample beat patterns and organise them in new ways as humans do. Also, virtually no birds move deliberately in time to an external beat, either on their own or in synchrony with others. The exceptions seem to be some members of the parrot family – or at least one member: a pet cockatoo named Snowball who can be seen online dancing more or less in time to 'Everybody' by The Backstreet Boys and 'Another One Bites the Dust' by Queen. New research also points to a hitherto unappreciated ability to synchronise to a beat among rats.

Humans have many limitations in comparison to other animals, but it seems we are uniquely able and adept when it comes to rhythm. Crickets entrain, or synchronise, with each other, but cannot sing; birds and whales sing, and vary their songs, but, with the apparent exception of cockatoos like Snowball, do not entrain. Humans do both, and create complex patterns that, as far as we know, no other beings manage. Our attraction to rhythm is such that we even search for and begin to pick it out in non-living natural and random phenomena. This is the effect in the piece 'Suikinkutsu – Water koto cave' by Stuart Chalmers, in which the sound of drops falling into buckets at different rates organises in the listener's mind into a kind of music.

Patterns where beats are grouped into twos, threes and fours are very common in Western music. A simple 'four to the floor' pulses at the centre of a masterpiece like Paul Simon's 'You Can Call Me Al'. But you don't have to go far to find others. In the Balkans and many points further east, music and dance uses metres with five, seven or eleven regular beats as a matter of course, as well as metres with irregular subdivisions. And in these and many other traditions, dancers and musicians play with, augment and push against the boundaries, creating almost endless variations that delight or heighten emotional intensity.

Still, there are only so many beats that we can count and

so many moves we can remember. This means that certain rhythmic patterns are universal in human music and dance. All include repetition, regular weak and strong beats, a finite number of beat patterns per song or dance move, and the use of those patterns to create motifs, or riffs (that is, repeated patterns or phases).

Nursery rhymes start simple. 'This is the way the ladies ride' is four beats. But things can quickly get more complicated. George and Ira Gershwin's 1924 hit 'Fascinating Rhythm' is structured in a way that is common to early-twentieth-century American popular music, with a sixteen-bar verse and thirty-two-bar refrain, but it endures in large part because of how it plays with this form. In the refrain, for instance (which begins 'Fascinating rhythm, you've got me on the go . . .'), an asymmetric figure is superimposed over a four-bar, sixteen-beat structure, creating two seven-beat phrases so that we jump – or, indeed, start a-hopping – to a syncopation that catches us with a cross-beat. The song can be crooned smoothly as it is by Ella Fitzgerald in her 1959 recording or expanded into a harmonic and pyrotechnical rainbow as Jacob Collier does in a version he recorded in 2014, but however it is packaged the effect is hard to resist.

The rhythms in Chaabi, a North African style associated with weddings and celebrations, are no less compelling for many who play, listen or dance to them. The groove in tracks like 'Amaliya' and 'Gwarir' by Karim Ziad, for instance, is a two-bar phrase with six beats per bar. Superimposed on this are high percussive notes on beat three of the first bar and beat two of the second bar, plus low percussive notes on beat five in both bars. The *qraqeb*, a castanet-like instrument made of iron, adds a further, wobbly rhythmic subdivision. The staggered pattern over the underlying regular drive gives the whole an enormous sense of energy and momentum. Musicians in many traditions also vary away from regularity by anticipating or falling behind the beat in order to create different effects. There are many other ways to explore rhythmical space, and express emotion with it. In contemporary jazz, and some other music forms, subtle changes in speed, in which beats are slightly delayed or accelerated in the context of a steady overall pulse, give the whole what the drummer Myele Manzanza described to me as the feeling of plasticine or a rugby ball, rather than the unyielding tautness of a football.

The Carnatic tradition of South India takes rhythmic complexity to an extreme. In Konnakol, a kind of spoken percussion, performers break down a regular time interval into anything from two to ten beats with a system of syllables that divide the whole in different ways. (One way to think of this is as a rhythmic counterpart to the Western *solfège* system, which uses different syllables to distinguish different pitches: where Julie Andrews as Maria in *The Sound of Music* has *Do*, *Re* and *Mi* to describe the first three notes in her scale, Konnakol has *Tha Ka* and *Tha Ki Ta* to describe a two- and three-beat rhythmic division of the same interval.) In partnership with others, South Indian musicians build astounding rhythmic structures. A tour de force of this kind of vocal percussion can be found online in a duet by V Shivapriya & BR Somashekar Jois.

Musicians and dancers will also vary tempo. A slow pace often expresses sadness, as it does in *Tristes apprêts* by Jean-Philippe Rameau, but it can also be meditative or tender as in the Adagio of Beethoven's Ninth Symphony. Fast tempos typically indicate greater arousal and happiness. Bulgarian folk dancing at its most exuberant can reach 520 beats per minute, or more than eight beats per second (approaching or even surpassing the 'tatum' – an interval named after the pianist Art Tatum for the smallest possible time interval between successive notes in a rhythmic phrase). But speed can express other emotions too. The Allegro in Shostakovich's Tenth Symphony, supposedly a portrait of Stalin, is ferocious and sarcastic or terrifying.

A speedy little number such as 'Bleed' by the extreme metal band Meshuggah may give a feeling of what the invasion of Ukraine in 2022 was like for some of the combatants.

A gradual increase in tempo is often associated with intensity, concentration and excitement. A DJ at a techno gig will increase the tempo over a sequence of songs, helping to create a mounting sense of ecstasy for dancers. And in some forms of *dhikr*, or devotion to Allah, Sufi worshippers begin to circle and chant together to a slow, steady pulse which gradually gets faster and faster. The rhythm of stamping feet is accentuated by the way the worshippers bend forwards to inhale on some words and stand straight to exhale on others. As momentum builds, the whole exerts a powerful effect on those taking part. 'Time unfolds, expands,' says one who appears in the short film *Tarikat* ('The Path'). 'A wave that will penetrate everything rumbles in the distance. Souls flow into the river of time.'

For all the centrality of rhythm to music and dance, there are works that seem to leave it behind. Ambient music such as Brian Eno's 1983 album *Apollo*, which contains few well-defined beats, seeks to lift the imagination literally into space. Melodies for the *guqin*, a plucked seven-string instrument that has been played in China since antiquity, are extremely free and open in their rhythms, and the music seems to expand beyond time as we usually experience it, as if in a realisation of the view that the closer a thing is to the Dao, or natural order of the universe, the less important time becomes.

All the same, humans seem to be irresistibly drawn back to rhythm. It awakens a deeply cooperative side of our nature. It signals group size, strength and coordination ability – how well we work together. 'When you impose rigorous order on musical rhythm, you are organising human motion,' writes the jazz musician Vijay Iyer. 'You create a dialogue between the physical and the ideal: embodied human action in a structured environment. The process gives us something to strive for, to work through, to achieve with virtuosity and grace.' In his recollection of David Mancuso, the 'psychedelic godfather of disco', Jeremy Gilbert recalls that Mancuso once told him that 'he often felt that all parties are just local expression of "one big party" taking place everywhere, all the time, that occasionally we manage to tune into or express through our own bodies and gatherings'. Gilbert, a cultural and political theorist, says that the profundity of this remark has stayed with him ever since. 'What [Mancuso] seemed to sum up in this single image was the fact that the joy of dancing in groups is an intense expression of the inherently creative capacity of the social relations that always constitute all of our being: what I call the "infinite relationality" of existence. The cosmic dance of matter, the multiplicity of the multitude, the creative power of complex groups: to acknowledge the god who dances is to acknowledge them all.'

Onomatopoeia

Bababadalgharaghtakamminarronnkonnbronntonnerronntuo
nnthunntrovarrhounawnskawntoohoohoordenenthurnuk!
grumbles the thunder in *Finnegans Wake*, and there is, as far as
I know, no longer example in English of onomatopoeia, or echo-
mimesis, in which the sound of a word is intended to recreate
the phenomenon to which it refers. The first of ten 100-letter
thunder-words in James Joyce's novel, it is also a miniature
map of humanity because, after the initial stuttering Babel of
'bababad', it is made from words for thunder in Arabic, Hindi,
Japanese, Italian, Irish and other languages.

Scholars point to tragic and serious allusions in this rattlebag
of a word. It represents, they say, the thunderclap associated
with the Fall of Adam and Eve, or the one which, according to
the eighteenth-century philosopher Giambattista Vico, terrified
early Man into taking refuge in caves, thereby giving rise to
language and civilisation. But there is no doubt that comedy
was part of Joyce's intent too. For him, the noise accompanying
the Fall of our first parents could also be the unheard noise of
a pratfall, as when Joyce's contemporary Buster Keaton walks
to the end of a plank projecting from the roof of a building
with the intent of jumping to the next, but instead falls down
through a series of awnings, grabbing a drainpipe which pivots

around and sends him shooting through a room on the floor below, where he slides down a fireman's pole only to be carried away on a truck.

Whatever Joyce was up to – and that discussion may never end – he was both a pioneer of this form of linguistic sound-play and late to the party. Among the masters of onomatopoeia are speakers of Taa, a Khoisan language spoken in South-west Africa by the Bushmen, or Sān. These hunter-gatherers are among the oldest surviving distinct peoples on Earth, having lived largely isolated from other humans for most of the last 100,000 years. They have some of the highest within-population genetic variance of any human group, indicative of that long isolation, and there is a sense in which their genetic diversity is echoed in their language, which has more different kinds of sounds in it than any other in the world. With five distinct clicks, multiple tones and vowels, Taa, which is also known as !Xoon, is reckoned to have as many as 164 consonants and forty-four vowels – or more than 200 distinct sounds. English, by contrast, has a total of about forty-five. This unusually large range enables speakers of Taa to be extraordinarily flexible and subtle in their spoken imitations of sounds. So, for example, a sharp object falling point-first into sand is ǂqùhm ǁhũ̀ũ, while a rotten egg being shaken is !húlu ts'ễễ, and grass being ripped up by a grazing animal is g|kx'àp.

Could onomatopoeia be the origin of language? Some early linguists thought so, and this notion, sometimes called the ding-dong hypothesis, has intuitive appeal. Certainly, imitating all kinds of sounds is an important part of language acquisition by young humans, just as mimesis plays a vital role for many young birds and whales. The presence and sophistication of onomat-opoeia in Taa would seem to support, or at least be consistent with, the case that it is among the longest-established parts of human language.

Onomatopoeia can edge into darkness with onomatomania:

an abnormal concentration on certain words and their supposed significance, or echolalia: the meaningless repetition of another person's spoken words that, in older children and adults, can be a symptom of mental illness. But for the most part it is a matter of fun and creativity, and it can spill over into plain bonkers inventiveness and delight in works such as the sound poem 'Ursonate' by Kurt Schwitters, which explores sounds on the borders of human vocal capability and which are seldom heard anywhere in the natural world. There is no doubt in my mind that, as Calvin and Hobbes observe, scientific progress goes 'Boink'.

As far as I know, every language has onomatopoeia, and there appear to be some near-universal characteristics. In tests, people from almost every background will tend to associate the made-up sounds *bouba* and *kiki* with, respectively, a rounded and a pointed shape. But we also enjoy learning how different languages represent the same natural sounds differently. A duck goes 'quack quack' in English but *coin coin* in French. In Spanish a dog goes *guau-guau*, not 'woof woof', while in Arabic it goes *haw haw*, and in Mandarin *wang-wang*. In Japanese cats go *nyaa*, and bees – having no access to the *zz* sound – go *boon-boon*.

Onomatopoeia is language at its least abstract – as close as it comes to the thing itself. An ultimate example, at the opposite extreme from Joyce's 100-letter thunder-word, must also be one of the shortest. The word that is written *Om* in English is pronounced 'aum', and according to the *Mandukya Upanishad* its first three phonemes express, respectively, the states of waking (a), dreaming (u), and deep sleep (m), while a silent fourth quarter denotes the infinite. In chanting the word, a speaker enacts what is believed to be the eternal emergence and return of sound in the world: the *Ātman* that is the essence, breath or soul of all.

How Language Began

In the beginning was the word. But what was the word in the beginning? There have been times when people have insisted that this question is not even worth asking. In a move that has since become notorious, the Société de Linguistique de Paris banned enquiries into the origins of language in 1866 on the grounds that there was too much speculation and too little hard evidence. But every age continues to have its ideas, and the present is no different. What has changed since 1866 is that there is considerably more evidence to go on.

Here is a sketch of what some who research the origins of language may agree is an educated best guess. It brings together updated versions of three older ideas or stories. The first of these – 'mimicry' – is the idea that human language started with onomatopoeia: the imitation of the sounds of non-human animals and other noises in the natural word. This seems intuitive, and it goes a long way back. Its champions have included the philosopher Giambattista Vico, who believed that the minds of the first humans resembled those of children today. Rather than naming objects conceptually, Vico said, early humans would have imitated those objects with monosyllabic cries – as well as with mute gestures. So, for example, when thunder struck, these first people would have imitated the shaking of the sky and shouted *pa* (father), creating the first word. It's a pretty fable, if a little silly by today's standards, but Vico was actually on to something: onomatopoeia remains a fundamental and enduring part of language.

The second idea – 'mime' – is a story about gesture, a human behaviour which is almost certainly very ancient indeed. Gorillas, with whom humans last shared an ancestor some 10 million years ago, use around a hundred different gestures in the wild, and almost nine-tenths of the fifty or so gestures commonly used by human toddlers are the same as those used by chimpanzees, with whom we last shared an ancestor about 6 million years ago. Bodily gesture and facial expression continue to play an essential role in many if not most face-to-face communications between sighted people today, and for those who use it routinely, sign language can be just as expressive and subtle as speech. A wholly new kind of language is emerging at present among the members of the DeafBlind community in the United States, who feel the gestures made by others with their hands. Protactile, as it is known, has an emerging lexicon of its own organised by new phonological rules, and even has a kind of 'tactile onomatopoeia', in which a hand resembles the feel of the thing it's describing by, for example, shaping a hand into a 'tree', with five fingers as branches, or a 'lollipop', with the fist as candy.

The third – 'music' – is the idea that human speech emerged from a kind of musical proto-language. Charles Darwin imagined this to be similar to birdsong: sounds that, he believed, had no specific meaning beyond being a way for males to impress females. At first sight there seems to be something to this. Rock stars of whatever gender often have sex appeal. But humans are not birds of paradise, where only the male is gorgeously ornamented. Women are no less capable musicians, and are often more articulate in speech than men. So perhaps, goes an amended version of this theory, the musical proto-language was for female-male duets. Or maybe it really began with exchanges between parents and infants in the songlike sounds of baby talk. Then again, it may have been for mutual protection: chanting in a chorus, it is argued, makes a group sound larger and more impressive to others, and will scare away dangerous predators.

If you thread mimicry, mime and music together to support each other, the resulting hypothesis goes something like this. Musical proto-language built trust – a feeling of 'us' – as it promoted release of endorphins, and enabled greater coordination between individuals. Also, with more complex rhythms and pitch patterns it became a signal to other human groups, as well as to animals, as to how well organised – and so how strong – the group was: a warning, or a welcome, and the origin of much of our music and dance. Learning and practising the use of the voice in new ways and in imitation of others would also support advances in communication during hunting and other activities. Still today, hunter-gatherers imitate the sounds of forest animals and birds to draw their prey towards them, and reproduce the calls of others so that they can locate different group members and coordinate their movements. Building on this, the practice would have helped establish the sense that a voiced sound can convey meaning – that it can manifest symbols for prey, friend and a hundred other things.

People would then have used many of these same sounds to accompany mimed performances or gestural storytelling. They would also begin to find that they could 'piggy-back' on the exhalation of breath to produce complex sounds that reinforce meanings and ultimately carry them on their own. Because speech 'rides' almost for free on the out-breath, it generally requires about ten times less energy than gesturing to transfer the same amount of information.

When did this all happen? The linguist Noam Chomsky has suggested that language as we know it developed over a few thousand to a few tens of thousands of years between about 200,000 and 60,000 years ago – that is, after the appearance of the first anatomically modern humans but before some of their descendants left Africa to populate the rest of the world. As language developed, it would have made possible the emergence of 'behavioural modernity' – complex activities that include symbolic culture and long-distance trade – which supposedly developed between about 100,000 and 50,000 years ago. The novelist Cormac McCarthy speculates that 100,000 would be a pretty good guess. 'It is fairly certain that [graphic] art preceded language,' he argues, although 'it probably didn't precede it by much,' and some of the oldest graphics found so far, in the Blombos Cave in South Africa, date to around that time.

But evidence discovered in recent years suggests that the roots of language go significantly further back. Anatomically modern humans lack the large air sacs in the throat which other great apes such as chimpanzees and gorillas use to make booming noises to scare off rivals, but which also inhibit the production of distinct vowel sounds that are crucial to more human-like speech. It is thought that early members of our genus such as *Homo habilis* still had these air sacs, but that they were not present in more recent members of our family tree. *Homo heidelbergensis*, which evolved after around 700,000 years ago from *antecessor*, the common ancestor of Neanderthals, Denisovans and ourselves, may however have possessed a hyoid bone suitable for supporting the tongue in movements that make possible subtle modulation of sound from the vocal tract. Further, both Neanderthals and modern humans appear to have inherited from our common ancestor a large number of nerve pathways from the brain to the diaphragm and the muscles between the ribs which enable the fine breath control that makes speech possible. Neanderthals also had a version of the FOXP2 gene, which influences the brain's wiring and plasticity in areas controlling speech, that was similar to (though perhaps less well adapted) than that of modern humans. What is more, the internal anatomy of their ears was also similar enough to that of modern humans to suggest they heard best in the registers typical of speech. All these lines of evidence, and others, suggest that speech of some kind had emerged by at least 765,000 to 550,000 years ago, when humans and Neanderthals diverged, and perhaps long before that.

Cormac McCarthy challenges the idea that language may be several hundred thousand years old. If that were true, he asks, what were humans doing with it all that time? 'What we do know,' he asserts, 'is that once you have language everything else [art, trade, complex social organisation] follows pretty quickly.' But I'm not sure this gets it quite right. Neanderthal societies may not have ticked all of McCarthy's boxes, but the complexity of their lives was commensurate with a role for language. In *Kindred*, a portrait of the Neanderthals and their world that takes account of the most recent discoveries, the archaeologist Rebecca Wragg Sykes shows that our cousins were far more capable and lived much richer, more complex lives than the knuckle-dragging caveman stereotype. Neanderthals made complex tools that required forward planning and multiple steps in processes such as glue-making. Speech would likely have played a role in conceiving and communicating such skills, and maintaining projects over space and time. Fear, pleasure, pain, excitement and desire would have flooded through their lives just as it does through ours, and it seems more than likely that they would have sought ways to share and communicate these feelings, as well as to organise and share knowledge. It also seems likely that language would have been part of this.

A little over ten years ago the cognitive scientist Deb Roy designed an experiment to observe and record the emergence of the first words spoken by his infant son. Cameras and microphones captured thousands of everyday interactions in the family home over a period of six months, and from this Roy, his partner and colleagues distilled a forty-second clip in which the child transitions from saying 'gaga' to 'water' when he wants a drink. It is a beautiful thing, and it inspires in me a wish that may be hard or impossible to fulfil – namely, to see a plausible recreation of the emergence of the first words spoken by our ancestors. This would be along the lines of the time-lapse that Roy made, only it would compress interactions across thousands or tens of thousands of years rather than a few months.

As for when to begin this epic time-lapse, I'd probably look to successive generations of our distant cousins *Homo erectus*, who lived and evolved from around 2 million to 100,000 years ago. Such evidence as has so far been recovered from this vast span of time suggests *erectus* were not capable of language as we know it. Certainly, their brains were smaller than those of Neanderthals and *Homo sapiens*. But we shouldn't underestimate them. Their bodies were as upright as ours, and sometimes taller and more robust, and over a period six to ten times as long as modern humans have existed they learned to control fire, develop complex stone tool technology, and hunt animals as big as straight-tusked elephants, which are larger than the African elephants alive today. There is evidence that they engraved symmetrical – perhaps symbolic – patterns, and were interested in ornamentation. Over tens of thousands of years they spread across Africa and Asia and probably built seagoing craft. Just imagine the gestures, voices and songs of these, our distant human ancestors, in a vast and stupendous world.

The Magic Flute

In 1971, on a visit to Lake Wallowa in north-east Oregon, a place sacred to the *Nimíipuu*, or Nez Percé tribe, the musician Bernie Krause followed his guide, an elder named Angus Wilson, up through a gorge towards the mountains. Gusts of wind funnelled through the gorge, and the walkers were suddenly engulfed by sounds that resembled those from a giant pipe organ. 'The effect wasn't a chord exactly,' Krause recalls, 'but rather a combination of tones, sighs, and midrange groans that played off each other, sometimes setting strange beats into resonance as they nearly matched one another in pitch. At the same time they created complex harmonic overtones, augmented by reverberations coming off the lake and the surrounding mountains.' Seeing that Krause and the other visitors were puzzled, Wilson picked up a few reeds of different length that had been broken off by the wind, and showed them how air flowing over the open holes at the top of them created the sound. He then took a knife from the sheath at his belt, selected and cut a length of reed, bored some holes and a notch into it, and began to play. Then, Krause continues, 'he turned to us and, in a measured voice, said: "Now you know where we got our music. And that's where you got yours, too."'

Cut from a plant or tree, a flute or a pipe (they differ mainly in the angle and shape of their mouthpieces) can seem to manifest the voice of a forest and, through echoes, a landscape. It can also make sounds that resemble the most musical of creatures, birds. 'Every day, instruments grow within forests and marshes,' writes the artist Ian Boyden. 'A wren perches on what might be a bassoon and sings . . . A reed-might-be-flute bends with a sparrow.' But while they can evoke or imitate nonhuman sounds, flutes and pipes can also marshal those sounds to human rhythms and emotions, summoning or evoking a sense of connection to the forest, the birds and beyond. These instruments can, in human imagination, give an audible form to what is felt to be invisible, primal and even divine: a voice for that which is beyond voice. So, for example, in the Yezidi tradition, it is said that the first soul refused to enter Adam without the accompaniment of the flute.

The oldest surviving purpose-built musical instrument discovered so far is a flute. Unearthed in a cave in the Swabian Jura in south-eastern Germany, it is made from the radius bone of a vulture and is about 42,000 years old. Modern replicas with five holes bored along their length and a V-shaped notch or blowhole on the end sound a pentatonic, or five-note, scale. In the 2010 film *Cave of Forgotten Dreams*, the archaeologist and Stone Age re-enactor Wulf Hein gives a rough idea of the instrument's range and timbre with a rendition of 'The Star-Spangled Banner' – though it could as easily have been 'Amazing Grace', 'Stairway to Heaven', 'Auld Lang Syne' or any number of other tunes that use this scale.

How bone flutes like this would actually have been played back in the Upper Palaeolithic is of course unknown, but a skilful player such as the musicologist Giuseppe Severini gives a sense of their fluency and potential in a clip you can find online. And in 2017 the flautist Anna Friederike Potengowski released an album of full-length re-imaginings. Consisting mainly of solo lines for the flute, with occasional accompaniment by simple percussion, tracks with titles such 'Wisdom 1' and 'Daybreak' are spare, meditative and intense.

Potengowski's reconstructions are just one possibility. There was almost certainly more to ancient flute-playing than this in the many different places where flutes may have been developed independently. The Bayaka pygmies who live in the Congo today play small wooden flutes. In recordings made by Louis Sarno, the players exchange looping phrases with fractional variations that carry far through the forest, rising above but also interleaving with the sounds of birds and insects. The effect is mesmeric – a kind of wild Ur-text for Steve Reich's composition for looped flutes, *Vermont Counterpoint*. Similar wooden flutes may have been played by ancestors of the Bayaka long before bone flutes appeared in Stone Age Europe.

The archaeologists David Graeber and David Wengrow argue that there is a tendency in our culture to underestimate the diversity and complexity of social and political systems which our distant ancestors created, and their willingness to experiment with new forms of organisation, including ones that are much less hierarchical than many people today typically imagine. And it may well be that our ancestors were often inventive and playful, as well as serious and devotional, in their culture and music. They 'knew where they stood in the scheme of things, which was not very high, and this seems to have made them laugh,' suggests Barbara Ehrenreich in a reflection on those who painted images of animals on the cave walls of Stone Age Europe. Alongside the rather earnest sounds summoned by Potengowski, there may have been times for something more like the Palaeolithic equivalent of Herbie Hancock's high-spirited 'Watermelon Man'. At other times flutes and pipes may have been tools used in hunting and associated activities, as Pascal Quignard imagines in *The Hatred of Music*: 'The small packs of humans who hunted, painted, and modelled animal forms would hum short phrases, execute music with the help of birdcalls, resonators, and flutes made from marrowbones, and dance their secret stories while wearing masks of prey as wild [*sic*] as themselves.' At times there may have been something of the sound made by the trickster and healer Treacle Walker as imagined by Alan Garner. 'It was a tune with wings,' writes Garner, 'trampling things, tightened strings, boggarts and bogles and brags on their feet; the man in the oak, sickness and fever, that set in long, lasting sleep the whole great world with the sweetness of sound the bone did play.'

What can be said for sure is that flutes were a small part of ancient material cultures. Figurines depicting mammoths, rhinos, tarpans and a human female dubbed the Venus of Hohle Fels were all discovered alongside the flute in the German caves. And vastly more is lost than survives, especially where objects were not made primarily of stone, bone or ivory. The few fragments discovered so far suggest that our ancestors made music with whatever they had to hand. Archaeologists have, for instance, found vestiges of a drum made from elephant hide dating from 37,000 years ago. Elsewhere, 24,000-year-old mammoth bones painted in red ochre bear marks suggesting they were struck repeatedly like a xylophone. A 15,000-year-old cave painting in France is thought to depict a musical bow – that is, a hunting bow adapted to be a musical instrument. Sets of elk teeth more than 8,000 years old found in north-west Russia have marks consistent with repeated rubbing made by the movement of dancers wearing them as rattling pendants.

Some musical instruments still in use also offer clues to sounds of the distant past. In the Serengeti, there is a rock gong that is almost identical to others across Africa and India that date to the Neolithic. Its stone slab makes different notes according to where it is struck, a little like a steelpan drum. Conches – large seashells that are punctured so that they can be blown like trumpets – found in India, Korea and across the Pacific today are virtually identical to a 17,000-year-old example found in a cave in France which was recently played again for the first time since it was lost. Shofars, often made of ram's horns, have been sounded in Jewish communities for thousands of years. They are almost certainly developed from older animal-horn instruments, possibly intended to frighten away evil spirits. The origins of the Aboriginal didgeridoo are unknown but, pre-carved as they are by termites, who eat out the heartwood from logs, these instruments were lying around all but ready to play when humans arrived in Australia about 50,000 years ago. In the Palaearctic, people may have tooted on mammoth tusks. 'If you put a hole in the side of one,' the archaeo-musicologist Barnaby Brown tells the writer Harry Sword, 'it produces the most amazingly intense drone.'

The *ney*, an end-blown flute which is typically cut from a giant reed but may also be carved from an eagle bone, has been played in Persia, Arabia and beyond for thousands of years. Alongside the Armenian *duduk* (which makes a sound by vibrating two reeds together in the manner of an oboe rather than by the passage of air over the edge of a hole), the *ney* is one of the most revered instruments of Greater West Asia that is still in use. For the poet, scholar and mystic Rūmī it was often an accompaniment to the Sufi dance which symbolised the soul emptied of desire and filled with a passion to return to God.

Side-blown or transverse flutes, where the player holds the pipe horizontally rather than vertically and blows across a hole in the side rather than one in the end, were common in India, Egypt, Greece and China by the second century BCE. In India, where they date back to at least the Indus Valley Civilisation of the third and second millennia BCE, these are called *bansuri*. They are usually made of bamboo and are still in use today. The *bansuri* is considered the divine instrument of Krishna, who played one when he courted the Gopis and their chief, Raisa, the goddess of love, tenderness and compassion.

European flutes became increasingly elaborate in the

eighteenth and nineteenth centuries, not least with the introduction of ring keys on longitudinal axles that brought more
distant holes within reach of the player. The newer models had
a more even tone and made possible rapid cascades of notes,
but tended to lack the sonic depth and timbral flexibility of
older versions. Even so, a sense of the instrument as a primal
force endures at the height of the classical period. In *Orfeo ed
Euridice* (1762), Christoph Gluck gives Orpheus his most beautiful melody for the flute in the underworld. And in *The Magic
Flute* (1791), in which Mozart sets a motley of fable, comedy and
drama to dazzling vocal lines and orchestral harmonies, it is a
simple wooden flute 'cut from the deepest roots of a thousand-
year-old oak' that enables the lovers Tamino and Pamina to
walk 'by the power of music in joy through death's dark night'.

In 'Ten Bulls', a series of pictures and poems in the Zen tradition, a practitioner of meditation who is advancing toward
enlightenment is shown playing a flute joyfully as they ride
the bull home, and in Japan the flute itself became a tool for
meditation. The *shakuhachi* is about as simple as it is possible
for a flute to be. Cut from a tall-growing species of bamboo
called *madake*, it has four finger holes on the front and one on
the back for the thumb. This is fewer than almost any other
common wind instrument, including the penny whistle. But in
skilful hands the *shakuhachi* can produce an astonishing range
of sounds, from breathy or harsh to pure and open, as well as
shifting microtones, flutters and other nuanced effects. Timbre
tends to take precedence over melodic line as the player focuses
their body, mind and breath towards *ichi-on-jo-butsu*, or 'enlightenment in a single tone'. A poem by the fifteenth-century monk
Ikkyū remains a guide for students of the instrument today:
'Playing the shakuhachi / One feels the unseen worlds / In all
the universe there is only this sound.'

From the eighteenth century onwards a repertoire for
the *shakuhachi* was codified, and public performance for

entertainment began. By the mid to late twentieth century the instrument attracted students and listeners from around the world, who esteemed it as a musical instrument as much as a tool for spiritual practice. One classic piece titled *Tsuru No Sugomori*, or crane's nest, was included on the Golden Record sent on the Voyager spacecraft.

The multi-instrumentalist Adrian Freedman, who grew up and trained in England but undertook years of study on the *shakuhachi* in Japan, has played it with everything from nightingales to tabla and cello, but returns to the solo instrument. He told me that the response he gets when playing it is quite different from when he plays Western instruments and styles. 'It resonates in a very deep place within the soul,' he said. 'After a *shakuhachi* performance I gave in São Paulo a couple of years ago a woman came up to me with tears in her eyes. "I am a social worker," she said. "My life is full of violence, but listening to your music, and the delicacy and subtlety of it, reminded me that I still have delicate and subtle parts inside me. I want to thank you for that." You could ask, what's the point of playing a bamboo flute in a world of such crisis? . . . I play as part of my own spiritual practice and because, in a small way, it helps give people strength and courage to go on.'

Explorations continue. *Glacier*, a 2010 piece for Western bass flute, is likened by its composer Dai Fujikura to 'a plume of cold air which is floating silently between the peaks of a very icy cold landscape, slowly but cutting like a knife'. And *Glacier* does sound like something from 'a different geological age', as the music critic Corinna da Fonseca-Wollheim writes: 'some notes splinter in two or dissolve into thin air, while, here and there, you can hear the ghost of a human voice channeled through the instrument'. In 2013 the flautist Claire Chase embarked on a twenty-four-year project to commission a new work for flute each year until the hundredth anniversary of *Density 21.5*, a 1936 composition by Edgard Varèse which Chase describes as one of the most concentrated four minutes in modern music. Works in the series so far combine the flute with electronic and visual effects to create new kinds of magic with one of the oldest musical instruments.

The Nature of Music

Nothing gets in the way of a good party for the Bayaka –
hunter-gatherers who live in the rainforest of northern Congo.
In a regular feature of camp life called spirit play, men and
women invite the forest spirits to join them in musical performances that include yodelling and polyphony accompanied by
percussive polyrhythms of clapping and drumming. To attract
the spirits out of the forest to play and dance with the humans,
the music must be beautifully performed, and the Bayaka give
it their all. From time to time, one of them will stand up to
dance and clown around, and when things are going just right
they may shout, 'Great joy of joys!' (*bisengo!*), 'Just like that!'
(*to bona!*), 'Again! Again!' (*bodi! bodi!*), 'Take it away!' (*tomba!*),
or 'Sing! Dance!' (*pia massana!*).

For the Bayaka, who have lived in much the same way for
tens if not hundreds of thousands of years, music is a powerful
force in almost every part of life. It begins at birth – or even
before, given that a foetus hears and feels its mother singing
and dancing almost every day while it is still in the womb.
Immersion continues as a baby listens to lullabies, dances
along on its mother's back, or sits in her lap while adults sing
together in a tight group. Music also has an important role in
hunting. When preparing for a hunt with nets, for example,

women will sing and play a flute to enchant the forest. They will sing late into the night and explain that this makes the animals feel *kwaana* – soft, relaxed and tired – so that they may be more easily caught. Men and women also perform music when they want to show off. The general principle, suggests the anthropologist Jerome Lewis, is that music and dance enchant sentient beings of all kinds, making them relaxed, happy and open.

It is hard to imagine music further away from the Bayaka and other pygmy peoples than the work of the avant-garde and experimental musicians of industrial modernity. But if you look into how some of these composers think about their craft, the distance is perhaps not so great. Iannis Xenakis, for instance, writes that music is the 'fixing in sound of imagined [worlds] . . . the gratuitous play of a child,' and George Crumb calls it 'a system of proportions in the service of a spiritual impulse'. These accounts describe the music of the rainforest dwellers quite well too. For all their playfulness, Bayaka songs and rhythms interlock to create complex, well-determined structures, even as they allow for individual improvisation and expression.

What is music? The *Oxford English Dictionary* defines it as 'the art concerned with the combination of sounds with a view to beauty of form and the expression of emotions'. The composer Edgard Varèse calls it 'organised sound'. Victoria Williamson, a music psychologist, describes it as 'a universal, human, dynamic, multi-purpose, sound signalling system'. Aniruddh Patel, a cognitive psychologist, terms it a 'transformative technology of the mind'. All of these do good service, but move us little further forwards. Music, like life, tends to elude definition.

One way to get beyond this is to think about numbers. The links between music and mathematics have long been recognised. In the classical and medieval eras in the West, music was one of four disciplines of the quadrivium at the centre of education. Music, as 'number in time', was studied alongside arithmetic ('number'), geometry ('number in space') and astronomy ('number in space and time'). 'Music,' wrote the natural philosopher Gottfried Wilhelm Leibniz at the turn of the seventeenth century, 'is the pleasure the human mind experiences from counting without being aware that it is counting.' Updating the definition for the YouTube generation, the musician Adam Neely suggests that music could be explained to space aliens as being based on the fact that humans are really good at hearing the relationships between regular events in time, and this capacity is at the heart of our perception of both rhythm and harmony. Music, he hazards, is 'the ritualistic application of mathematical ratios. If mathematics is a constant throughout the universe so too is the concept of human music.'

But music is almost always a lot more than this. For Arthur Schopenhauer, writing in the early nineteenth century, music was the direct expression of what he called the Will: a desire, striving and wanting which he believed to be the innermost truth of the world. You don't have to go that far, however, to recognise that music shares key features with, or at least mimics, motions and emotions that are essential to the experience of being alive – and the endless variations thereof. To make or listen to music requires physical movement and/or physiological arousal as well as the stimulation of neurons in more regions of the brain than almost any other activity, linking deeply into areas devoted to vision, motor control, emotion, speech, memory, planning and sexuality. 'What might be special about music,' writes the musician and psychologist Elizabeth Hellmuth Margulis, 'is not so much that it is different from everything else, but rather that it draws everything else together.' If consciousness is what happens when our minds interpret our bodies and feelings and reflect their processes back to us, then music is to some extent a means of rehearsing, reflecting upon and exploring the permutations and possibilities of existence and emotion. Like consciousness, music presents models of life, which it sometimes seems to resemble so closely that we almost take it for life itself.

Crucially, music helps people find and define their place in the world, and connect with others. 'Each social group has its music,' observes the musicologist Bruno Nettl, and the music each of us treasures connects us to that group. Intensity in music, writes Chris King, the author of a study of the ancient folk music of Epirus in north-east Greece, 'is a result of inter-connection between a place, its people, and its music'. Music is, then, part of a set of practices and knowledge that brings a community together and enhances a sense of identity. It is a shared voice.

And music is also a way of interacting with the larger-than-human world. It builds on 2 million years or more in which the genus *Homo* has been paying attention to the sounds of nature. Some parts of that world-song, such as the performances of the pied butcherbird, an Australian fowl of astounding vocal agility, may never be equalled by humans – although Thelonious Monk and Olivier Messiaen did not stop trying.

Further, music connects people to whatever they apprehend or imagine beyond the world they see and observe. The drone, or continuing underlying tone, in Indian classical music is sometimes regarded as a manifestation of *Nadha Brahma*, a term from the Vedas that translates as the 'sound of God': the vibration that runs through everything. And according to the eleventh-century Muslim philosopher and theologian Al-Ghazali, 'what causes mystical states to appear in the heart when listening to music [*Sama*] is a divine mystery found within the concordant relationship of measured tones [of music] to the [human] spirit'.

Music, wrote Rainer Maria Rilke, is 'language where language ends . . . the innermost point in us [that] stands outside'. For John Cage it is 'a bubble on the surface of silence'. Perhaps, as the bubbles rise and pop, we hear eternity in love with the productions of time.

Harmony

Sometimes a good thing happens when you are singing a part song with others and the intervals between your different notes come out just right. As the harmonies lock into place, the sound seems to become brighter, richer and fuller. It is as if the spaces between singers are filled with warmth and brilliance, and your body seems to glow as it resonates. The writer Diane Ackerman likens the effect to having a massage on the inside. Boundaries evaporate so that, as David Hume said in another context, 'all the affections readily pass from one person to another, and beget correspondent movements in every human creature'.

A simple definition of harmony is that it is a state of being in agreement or concord. In music, it is a combination of simultaneously sounded notes that is (usually) intended to produce a pleasing effect. The word derives from the Greek ἁρμόζω, or *harmozō*: to fit together, or join. Its analogue in Sanskrit is 'yoga', which also means to join or unite: the union of the individual, *jīvātmā*, with the supreme self, *paramātmā*. For Confucius, 'harmony is ... the unification of plurality. Music is the harmonisation of heaven and earth. Where there is harmony the myriad living things are in accord.'

Music is no one thing. In many cultures it is not separate from dance. But while the harmonies that people find pleasing vary across cultures and time, some aspects of it are close to universal. Foremost among these are the octave and the fifth – the intervals between, respectively, the first and second notes in 'Somewhere Over the Rainbow', and the notes for 'Baa Baa' and 'Black Sheep'. And these two can be a ground for endless form and invention. In Indian classical music, for instance, an instrument such as the *tanpura* will typically drone (make a continuous sound) on the fundamental (that is, the home note) and the octave and fifth, setting an underlying tonality for the piece while microtonal and rhythmic variations unfold over time.

The octave and the fifth are produced by sound waves that bear simple physical and mathematical relationships to the fundamental. The octave is made by a wave half the length of the one that produces the fundamental, the fifth by one that is two-thirds as long. Expressed as ratios, they use the smallest natural numbers – 2:1 for the octave, and 3:2 for the fifth. In the system known as just, or pure, intonation, other intervals in the musical scale are also ratios of small, whole numbers. A fourth is made by a ratio of 4:3, a third by 5:4, and so on.

Because the ratios of these waves are so simple, their peaks and troughs come in and out of synch with each other in regular, predictable ways. The octave makes two vibrations in the same amount of time as the fundamental makes one, while the fifth makes three in the time it makes two. The human ear and brain recognise and welcome these simple relationships rather as when, standing in a boat, we might welcome regular, rolling waves but not a choppy, uneven sea. This may go some way towards explaining the physical basis of the experience of harmony.

Harmony is also enfolded within every note. Again, this comes down to physics and maths. When a string on a musical instrument such as a guitar or the column of air inside a flute – or the human respiratory tract – is agitated, it creates a wave of sound as long as the string or air column. This produces the fundamental: the note we hear most prominently. But it also vibrates with shorter waves that make up what is known as the harmonic series. This is a set of higher pitches 'stacked' on top of each other in an octave, a fifth, fourth and a sequence of ever smaller and higher intervals in something like a musical rainbow. Most of us, most of the time, do not hear these overtones distinctly, but we do sense the timbre, or 'feel', that they and other less consonant overtones create. How loud these overtones are in relation to the fundamental is a vital part of what makes a note played on a harp sound different from the same note played on a trombone – or a rubber band stretched over a shoebox. Harmony is also linked to melody and rhythm in profound ways. A melody is a series of harmonically related notes sequenced across time rather than sounded all at once; and any given harmonic interval is mathematically identical to two different rhythms played against each other.

One way of telling the story of music in Europe is as a gradual exploration of harmonic space. For the first thousand years and more of the Christian Era, a lot of it seems to have been monophonic – that is, a single, simple line such as we still hear in Gregorian chant. Harmony was, as far as we know, limited to two forms, at least in church. With *organum*, named for its supposed resemblance to the sound of an organ, voices or instruments sang or played a second note in lock-step with the tune, but an octave, a fifth or a fourth apart. With a drone, a single note was sustained without change as the chant unfolded above. Compositions by Kassia, a ninth-century Byzantine abbess who set her hymns to elaborate melodic lines, are striking examples of this style, as are the works of her twelfth-century German counterpart Hildegard of Bingen, who created remarkably free, wide-ranging chants which she called celestial harmonic revelations in sound.

The next steps were taken in Paris a generation after Hildegard, when Pérotin, a composer of the Notre Dame school, combined as many as four different melodic lines to create dense harmonic textures that, as far as we know, had never been heard before. He also used sprung rhythmic patterns. For all his innovations, however, Pérotin continued to use the intervals of octave, fifth and fourth, and his harmonies sound austere to modern ears. In the following two centuries musicians also increasingly deployed thirds and sixths. Guillaume de Machaut used thirds in his *Messe de Nostre Dame*, which he composed sometime before 1365. But he did so mostly in passing, unaccented moments, and the third, which is the interval between the first two notes of 'Kumbaya', only began to take a central place in Western music in the fifteenth century with the works of John Dunstaple and Josquin des Prez.

The results have been likened to a chemical reaction, and even a miracle cure. Sandwiched between the fundamental (or first note) and the fifth, the third co-creates the rich, thick chord known as the triad. By the early seventeenth century Claudio Monteverdi was using triads to magnificent effect in works such as his *Vespro della Beata Virgine* of 1610. Musicians have continued to experiment with harmony in countless ways ever since, but few have abandoned the triad completely, or for long. The conservative nineteenth-century music theorist Heinrich Schenker called it the 'chord of nature', and argued that it was at the heart of every great piece of music. Still today it has a place in some of even the most outlandish musical journeys. Jacob Collier, a musician noted for bold harmonic experiments, calls it 'almost like acoustic truth'.

Thirds and triads, and harmonies that build upon them, have been a vital part of the European tradition for hundreds of years now. They often play an important role in harmonic progression in which a sequence of chords moves away from the key based on a home note, or 'tonic', before returning to it at the end. This typically creates a sense of resolution – of coming back home – in what is known as a cadence. Particularly common is a cadence from a key based on the fifth, or 'dominant', back to one based on the tonic, with which it shares all but one note. This is called a perfect cadence – a five one, or, in Roman numerals, V, I – and it can be found everywhere from Beethoven's Fifth Symphony to 'I Want to Hold Your Hand' by The Beatles, where it comes at the end of every chorus. Another common cadence, called the plagal, resolves from four to one (IV, I). It is sometimes called the Amen cadence because it's how many hymns end, but it also features in songs like Abba's 'Mamma Mia' (in the verse at 'I've been cheated by you', and in the chorus at 'My my'). Some progressions have become so agreeable to Western ears that they can be hard to escape. In 'Four Chords', the musical comedy act Axis of Awesome satirises the lack of harmonic variety in dozens of global hits, from 'Let It Be' and 'No Woman No Cry' to 'With Or Without You' and 'Can You Feel the Love Tonight', by showing how they all follow the same I–V–vi–IV progression.

Thirds come in two forms: major and minor. Those of us who have grown up in the Western tradition tend to think of the first, which is made up of four semitones, as bright or happy, and the second, which is made of three semitones, as dark or sad. And it may be that these perceptions are to some extent rooted in the embodied origins of music. Brightness is defined, simply, by the larger relative size of the intervals between notes, and this may be linked to posture in the human body. We stand upright and tall when dancing or moving with joy and confidence, and curl or hunch when experiencing an inward or darker turn. But whatever the origin of our feelings for major and minor, the two exist in tension, and are seldom free of each other. Harmonically, major and minor triads are mirror images of each other: in the first, a minor third is placed on top of a major third, while in a minor triad it is the other way round.

But harmonic intervals can interact to produce effects which seem to express flows and states of feeling that are more complex and subtle than just happy or sad. Consider the Prelude in C major from Book 1 of Bach's *The Well-Tempered Clavier*, which was first published in 1722. The entire piece is a sequence of arpeggios – that is, broken chords – built upon thirds. The first establishes a C major harmony, with a C followed by an E natural. The second gives us D minor, with a D (over C) followed by an F natural. In this way, as the musician and theorist Adam Ockelford observes, a motif that is at first positive and cheerful is almost immediately transformed into something darker or sadder. This change, however, is moderated by the fact that, in a move that is generally felt to give a positive sense of change, the pitch of the top line rises. At the same time, the lowest note, the held-over C, creates a dissonance that has a sense of needing to be resolved. 'This intricate fusion of similarity and difference, of becoming major and minor, imbues the music with a sense of yearning,' writes Ockelford. It gives rise to 'a complex emotion that, amazingly, is evoked by a series of abstract sounds through the blend of regularity and irregularity inherent in the design of the major scale'.

Harmony always exists in relation to dissonance. Monteverdi used clashes and dissonances in his madrigals and operas to express pain, sorrow, anger and other emotions, and many have followed in his path. Harmony can also, through unfamiliar modulations, take us to places that can be hard to label: the dark-and-light in Franz Schubert's late string quartets and piano sonatas, or a new colour palette in John Coltrane's *Giant Steps*. That said, not all harmonic journeys are subtle or elusive. In what is known as the truck driver shunt, the music unceremoniously grinds up a gear – that is, one key. A good example occurs at about one minute forty-five seconds into 'My Girl' by The Temptations, and also about three minutes into Whitney Houston's version of 'I Will Always Love You' just after the deep drum 'booj'. The result for the listener is a near-guaranteed sense of elevation and euphoria. But there are other, less predictable, ways to get there. For Jacob Collier, a version of 'A Quiet Place' by the group Take 6 has towards its end his 'favourite modulation of all time' as the harmony lifts and lifts and bursts forth. 'I can so clearly remember discovering harmony in this way,' he recalls, 'thinking I did not know it was possible to achieve those emotions just with notes in chords.'

Harmonies made with notes on the plain-vanilla, seven-note scales typical of European classical music and much of rock and pop are a small part of the story of harmony as a whole. Pentatonic scales, with just five notes per octave, are common across the world, from the Andes to the British Isles, though they come in many forms. In Ethiopia alone there are at least four different pentatonic scales, known as *kignits* or *qenets*, while a five-note scale that consists of the tonic, the major second, the minor third, the major fifth and the minor sixth seems to transport the listener to Japan. Blues scales typically add a flat five, which may originate in West African music, to a pentatonic scale to make a six-note scale.

Western music divides an octave into twelve semitones, and typically uses seven of these twelve in a given key, but there is no reason to limit the number of divisions in an octave to twelve. Continuing the series of whole-number ratios that yields the perfect fifth, fourth and so on results in nineteen distinct tones, and before the adoption of tempered tuning in Europe in the seventeenth and eighteenth centuries many musicians would have been accustomed to these gradations – distinguishing, say, an A sharp from a B flat where we are now used to something in between. Other musical cultures continue to glory in microtonal melodies and harmonies. Indian classical music slides across microtonal ranges, and Arabic music typically divides a scale into twenty-four quarter-tones.

We can only begin to imagine the harmonic worlds that may come into being in the future. Will they be variations on sounds we know and love, or take music into unexplored territory?

Compositions by twentieth-century rebels such as Harry Partch and Karlheinz Stockhausen tend to have limited appeal today. This suggests that much of what humans enjoy will not change very much. But then again, the legacy of those who explore the unknown may be more fertile than is readily apparent. There are strange new resonances in the work of Radiohead, Björk and many others.

Maybe Stockhausen's *Gesang der Jünglinge*, which mixes the human voice and electronics in ways that, in 1956, had not been heard before, presages a world in which humans are increasingly joined to computers. Maybe we will hear more and more harmonies that are unlike those we have known in the West for the last few centuries. Maybe they will be informed by ever-greater attentiveness to sounds of the non-human world. As technology advances, so will our ability to hear and tune into the music of other species. For now, one can listen to a long-distance call of the blue whale, which has only recently been discovered and recorded. Sped up to at least twice its natural speed in order to be audible to humans, the 'z-call' repeatedly swoops down an interval somewhere between a perfect fifth and a tritone. An end and a beginning.

Strange Musical Instruments

'Madam!' exclaimed the conductor Sir Thomas Beecham, 'You have between your legs an instrument capable of giving pleasure to thousands, and all you can do is scratch it!' He was, of course, referring to her cello. And it's true that this instrument, which has not changed much since it was first developed around 500 years ago, is capable of a beauty and richness of tone that is hard to beat. This has not stopped people from trying. There may be only around 300 different kinds of pasta, but there seem to be almost no end of possibilities in the realm of what a librettist for Henry Purcell called the 'wondrous machines' of music.

A starting point for finding one's way in the landscape of all conceivable musical instruments is a system of classification that divides them into five main groups. Devised by Erich Moritz von Hornbostel and Curt Sachs in 1914, it initially had four: idiophones, which make a sound by vibrating all or most of their bodies (bells, gongs, marimbas); membranophones, which vibrate a membrane (mainly drums); chordophones, which vibrate strings (sitars, koras, violins, guzhengs, pianos); and aerophones, which vibrate columns of air (pipes, woodwind, brass, organs). For almost all of human history this would have been enough, but the harnessing of electromagnetism in the twentieth century led to an expansion of possibilities that continues to this day, and in 1940 Sachs added a fifth group – electrophones – to describe them.

There are any number of oddball musical instruments in any of these groups once you start to look. There are piccolos as small as pencil stubs, and subcontrabass flutes that look like the giant pipes you might find beneath storm drains in the kingdom of the upside down as conceived by a committee that included Gerard Hoffnung and Hieronymus Bosch. There is the kangling, a trumpet made from a human leg bone, gifted to the world by Buddhists who wanted us to make mortality audible. There are members of the violin family called octobasses which are as tall as a short-faced bear – that is, twice the height of a man. And there is no end of tuned and untuned percussion made out of plastic pots, PVC pipes and other gubbins. Sometimes all you need to do to invent a new musical instrument is to rename an old one, as Sun Ra did when he prefixed 'space-dimension' to mellophone, called an electronic keyboard a 'cosmic tone organ' and a piano a 'sunharp'.

And then there are the chimeras: contraptions that bring together what many of us think of as two different instruments to make something new. A violin with a keyboard never caught on, and probably for good reason. But the gameleste – a combination of celeste and gamelan created for Björk and featured on her track 'Crystalline' – works pretty well. The harp guitar, which was first developed in the nineteenth century, is enjoying a modest resurgence. You can see dozens of players across North America and Europe joining in a virtual performance of 'The Water is Wide' and hear it on *Urban Driftwood* by Yasmin Williams. With twenty or more strings mounted on two necks, one of which is also a curved sounding board joined to the main body, the instrument is something like a cubist flamingo crossed with a theorbo (a giant Renaissance lute which also has two sets of strings). But among the strangest, devised back in the seventeenth century and possibly a myth, is a cat piano, which yanked the tails of different cats tied to different keys on a keyboard, causing them to screech. This macabre invention was reimagined in benign form a few years ago by the musical inventor Henry Dagg as a rack of plush toy cats, each of which squeaked a different note when squashed for a rendition of 'Somewhere Over the Rainbow'.

Musical instruments that are partly or completely automated have a long history. Organs powered by water date back to at least ancient Greece, and may have been used to imitate birdsong as well as to play music. In *The Book of Ingenious Devices* published in Baghdad in 850 CE, the Banū Mūsā brothers describe a version which played interchangeable cylinders automatically, as well as a steam-powered flute. Some commentators have claimed the former as the first programmable machine. Mechanical bell-ringers controlled by rotating cylinders were being made in Flanders by the fourteenth century. They were the distant ancestor of the music box developed around 1800, and which appears to such dramatic effect at the climax of the 1965 film *For a Few Dollars More* as it gradually winds down during the climactic confrontation.

Increasingly elaborate and complex methods of sound production and amplification were a subject of fascination at least as far back as the seventeenth century. By around 1615, the engineer Salomon de Caus (drawing on the work of Philo of Byzantium and Hero of Alexandria in antiquity) perfected a device which used sunlight to heat a copper vessel in such a way that water was pushed up a pipe to make fountains flow, and statues to play music. In *New Atlantis*, a utopian novel published in 1626, Francis Bacon imagines 'Sound Houses, where [they] practise and demonstrate all Sounds, and their Generation' and reproduce 'the Voices and Notes of Beasts and Birds' and have a 'diverse strange and Artificial Ecchoes Reflecting the Voice many times, and as it were tossing it'. And in his *Musurgia Universalis* of 1650, Athanasius Kircher revived the aeolian harp of classical antiquity – an instrument that relies on the wind to vibrate strings over a sounding board. Inspired by these and other concepts, John Evelyn describes in his *Elysium Britannicum* of 1700 a range of artificial constructions, and 'wonderfull' musical garden automata. These include 'an Æolique chamber' that imitates the chirruping birds; a Watchman that sounds a single trumpet note; a speaking statue of Memnon; and (in what sounds like a descendant of an invention described by the Mūsā brothers) an Autophône Organe, which will play whatever musical piece is 'deliniated' on its 'Phonotactique Cylinder'.

Over the course of the nineteenth century musical automata increased in sophistication, and sometimes in wackiness. In addition to the fairground organ and the pianola that are still relatively well known today there were such marvels as the Belloneon, which could play twenty-four trumpets and two drums, and the Panharmonicon, which could imitate most orchestral instruments as well as gunfire and cannon shots. The Orchestrion went one better by incorporating a piano. The Componium could play endless variations on a theme fed into it. The Apollonicon tooted on 1,900 pipes.

Even in modern times, when almost any sound can be synthesised electronically, ingenious mechanical instruments continue to delight. The Great Stalacpipe Organ built in the 1950s in the Luray Caverns in Virginia in the USA bumps little rubber mallets against stalactites. A litho- (rock) idiophone three and a half acres in extent, it may be the world's largest musical instrument. The Marble Machine completed by the folktronica band Wintergatan in 2016 uses a hand-crank to lift steel marbles and feed them into tubes, whence they issue through programmable release gates to fall and strike a vibraphone, bass guitar, cymbal, high hat, kick drum and snare drum. A video of the machine in operation has been viewed more than 225 million times at the time of going to press.

'Musical sound is too limited,' wrote the futurist Luigi Russolo in his 1913 manifesto *The Art of Noises*. In an age of industry, speed and clamour, he said, an 'infinite variety of noise-sound' must be 'conquered' because 'we find far more enjoyment in the combination of the noises of trams, backfiring motors, carriages and bawling crowds than in [Beethoven]'. The instruments of the future, he suggested, must be capable of thunderings, explosions, roars, bangs, booms, whistling, hissing, puffing, whispers, murmurs, mumbling, muttering, gurgling, screeching, creaking, rustling, buzzing, crackling and scraping, not to mention shouts, screams, shrieks, wails, hoots, howls, death rattles and sobs.

To create these noises, Russolo designed and built new instruments which he called *intonarumori*, or noise-players. These were, essentially, large wooden boxes mounted with amplifying funnels and external levers that moved mechanical arms inside to strike, scrape and vibrate against drums, wires and other devices. These simple acoustic contraptions were, Russolo believed, just a start. 'Tomorrow, as new machines multiply, we will be able to distinguish ten, twenty, or thirty thousand different noises [and] combine them according to our imagination.' And to an extent his vision has been realised over the century that followed in *musique concrète*, noise, industrial, sound art and more.

Some of those musical experiments would be hard to repeat. Take Arseny Avraamov's *Symphony of Factory Sirens* that took place in Baku on the fifth anniversary of the Soviet Republic in November 1922. This deployed several large choirs, the foghorns of the Soviet Caspian flotilla, two batteries of artillery guns, several infantry regiments, including a machine-gun division, hydroplanes, and all the factory sirens of the city. Conductors posted on specially built towers signalled various sound units with coloured flags and pistol shots. A central 'steam-whistle machine' pounded out 'The Internationale' and

'La Marseillaise' while 'autotransports' raced across Baku for a gigantic sound finale in the festival square.

Other musical novelties had a more domestic flavour, but might prove difficult to recreate at home. There is, for instance, Filippo Tommaso Marinetti and Luigi Colombo Fillia's *Manifesto for Futurist Cooking* of 1930, in which knives, forks and politics are abolished but sound plays a central role. For Dish 1, 'the Aeropainter', guests eat kumquats while stroking sandpaper and having aircraft noise blared in their ears. For another meal the first course is a 'polyrhythmic salad', consisting of a box containing a bowl of undressed lettuce leaves, dates and grapes. 'The box has a crank on the left side. The guests eat with their right hand while turning the crank with their left. This produces music to which the waiters dance until the course is finished.'

From Theremin to Moog, and from live electronics to laptronica, the range of sounds that it is now possible to make has grown almost without limit. And as it becomes easier to do so electronically one might almost wonder if musical instruments will be replaced entirely by software. And yet, we continue to cleave to the physical objects. Humans thrive when we are in touch and communion with the material world, and we seem to be compelled to keep making music with instruments we can handle and hold, squeeze and strum, stroke and strike, bow and blow.

Which among a thousand twangling instruments will endure? It is likely that some of those that are already long established will continue, just as (other things being equal) animal and plant species that have well-established ecological niches tend to survive. This may be especially true for some of the older instruments that can make new kinds of sounds as well as those for which they are well known. Consider how the violin explores microtonal and microtimbral space in a piece like *Birds in Warped Time II* by Somei Satoh, and experiments by Mari Kimura in which she has revealed a world of subharmonics on the violin that were previously believed to be impossible. And perhaps other old instruments will find subtly new forms. One example may be the Cristal Baschet – an 'organ' made of chromatically tuned glass rods that are gently stroked to make a sound. Played for or by Ravi Shankar, Damon Albarn, Daft Punk, Radiohead, Tom Waits and others, it is essentially an update of a musical instrument called a glasspiel or verrillon, which is created by filling glasses with varying amounts of water, and which dates back to at least the early eighteenth century.

The future for early electrophones is less clear. When the Theremin, a device in which the musician varies a wobbly note by moving their hands with respect to an antenna, first appeared in the early twentieth century it seemed like the wave of the future, but it now looks increasingly like a curiosity. By contrast, the Ondes Martenot, which is a kind of electronic keyboard, shows signs of revival, thanks to recent interest from musicians such as Jonny Greenwood. Perhaps the future is hybrid: instruments that arise from the interaction of electronic system with their human creator. Daphne Oram's Oramics Machine, conceived and designed between 1962 and 1969, which synthesises and sequences sound from lines and marks that she painted on glass and film, may be an early example. In 2010 Imogen Heap began developing Mi.Mu gloves, a 'wearable musical instrument for expressive creation, composition and performance', which take this kind of creativity into three dimensions and dance in real time, as the musician moves their gloved hands to make new sounds. It may be a long shot but there's also the Segulharpa, an electromagnetic harp designed by Úlfur Hansson, which looks like an artefact from an alien civilisation imagined by Ursula Le Guin. 'Being able to control a physical object electronically opens up new possibilities in the way you interact with spaces and listening,' says Hansson. The

Segulharpa is 'always evolving as you play. You can feel that it's shaping itself.' Alternatively, electronic instruments may bring the living world into their 'heart'. To take a madcap example, Cosmo Sheldrake's *Pliocene* (2018) is composed with samples of sounds of animals in endangered ecosystems. So it is serious, but also – with a kick drum that is an oyster toad fish, and a buck-toothed parrot fish for a snare drum – zany. Sheldrake first performed it while floating over Barcelona in a hot air balloon.

Even if few new instruments endure, there will almost certainly continue to be an endless stream of experimentation and invention. I hold out hope for creations such as the Daxophone, a friction idiophone made of wood that produces an astonishing range of throaty, often comical, voice-like sounds. And there are any number of others. Some, such as Constance Demby's giant sheet metal idiophones called the Whale Sail and the Space Bass, which are played with a bass bow to create low resonant tones, seem to try to transcend the planet. Others are particular to a time and place. The *útiles sonoros* – sound tools – created by the composer Joaquín Orellana include 'beyond-marimbas', which resemble toy roller-coasters or surrealist sculptures as much as they do marimbas. With these, Orellana conjures sounds that speak of both the joyous music of his native Guatemala and the terrible suffering of the country's long civil war. Elsewhere, some of the most beautiful new sounds being created today may be those made by new kinds of steel drums, or pans, such as those being created by Kyle Dunleavy for Josh Quillen of Sō Percussion. These create a 'teardrop' sound – 'really bright at first, and then there's a decay, a spreading out and a softening,' according to Quillen. 'The sound is dark, but there's a point to it . . . [it seems] to take on the quality of a human voice.'

Some instrument-makers and musicians make a point of working, almost like Foley artists, with materials that are ephemeral or to hand. The Vegetable Orchestra of Vienna create an amazing salad of rhythmic and harmonious sounds with gourd drums, carrot pipes, leek oboes and more. They make new instruments for every show, and turn them into soup afterwards. Every year, the Ice Music Festival in Norway features a new instrument made of . . . yes, ice. Star turns have included an ice-didgeridoo, an ice-harp, an ice-udu (a percussion pot that makes a range of tones), an ice-balafon (like a xylophone but with gourds for resonance), and an ice-kantele (a plucked string instrument of the zither family with a bell-like sound). These instruments are closely modelled on their conventional counterparts, but differ in vital respects. A double bass made of ice, for instance, is denser and heavier than the wooden original. 'Harmonies must be simplified and played more slowly, requiring improvisation and a renewed state of mind,' says the bass-player Viktor Reuter. Taken as a whole, the festival is a meditation on nature and, because the instruments will all melt away, on climate change.

Each of the five groups of musical instruments in the Hornbostel-Sachs system contains dozens and even hundreds of categories, and within each of the categories there are hundreds and even thousands of distinct instruments – past, passing and to come. Even so, all the instruments in all the categories – with the partial exception of electrophones, which do much of their work inside a black box – echo our physical nature in the world. We ourselves are idiophones when we stamp in resonant space, membranophones when we beat our chests, chordophones in that our vocal cords vibrate in a similar way to strings, and aerophones when we shape voice and song in our chests, throats and mouths.

For hundreds of thousands of years, our ancestors have heard sounds in the world that they interpreted as music. For a good part of that time, if not all of it, they sought to join in. Astonishment and delight must have been frequent guests during the discovery and development of the first musical instruments: stones that produced bell-like tones when struck, pipes made from the bones of birds, and conch-shell trumpets. These instruments not only extended human capabilities but also seemed to summon voices and songs from the non-human or more-than-human worlds. With the best of the strange musical instruments being developed today perhaps we experience something of the same emotions – revelation, glee, solace, fury, puzzlement – as our distant ancestors. And just as some of the oldest instruments endure alongside the youngest, so, perhaps, our descendants will continue to marvel and enjoy both the old and the new.

Sad Songs

Humans are not alone in making sounds and gestures that express sadness. Elephants vocalise in distress when one of their bond group is dying, and will stand guard over the beloved dead for many hours, moving quietly and slowly, and sometimes touching the body tenderly with their trunks. Wolves howl for dead companions in a way that humans familiar with their normal behaviour describe as exceptionally soulful and heart-wrenching. Our ancestors would have observed such behaviours in non-human animals over hundreds of thousands of years, and been influenced by them. And, as we featherless bipeds became skilful in the use of our own voices, song would have played a major role in the expression and sharing of pain and loss, as well as love and joy.

We can imagine what the sad songs of early humans would have sounded like and how they may have changed over this vast expanse of time. But we only have direct evidence in the written record for the last tiny fraction. Take the Psalms. These Hebrew sacred songs were assembled between the ninth and fifth century BCE, but drew on earlier Canaanite, Egyptian and other sources. The Hebrew term for psalm is *mizmor*, which means 'something sung', while the collection of 150 that has come down to us is known as *Tehillim*, which means 'praises'.

Most of them are hymns of praise, but a good number are laments: vivid expressions of suffering and despair. 'I have sunk in the slime of the deep, and there is no place to stand,' goes Psalm 69. 'I have entered the watery depths, and the current has swept me away, / I am exhausted from calling out. My throat is hoarse.' Words like these have spoken powerfully to succeeding generations, and have inspired music of great beauty in recent centuries, but hints of their original sound survives, too, in the tradition of cantillation, the voicing halfway between melody and recitation that is notated in a system of accents (ta'amim), or diacritical marks, in the written text.

There are also glimpses from ancient Greece. One example is a fragment of the chorus in Euripides' tragedy *Orestes*, which was first performed in 408 BCE. The words read: 'I grieve for you – how I grieve for you. Among mortal men great prosperity never lasts. Some higher spirit shatters it like the sail on a fast ship and hurls it into waves of dreadful sorrow, as deadly as storm waves out at sea.' Marks on the text tell the singers to begin on what today we call B flat, descend by a microtone, and then continue down to A natural. Meanwhile, a drone, or continuous single note, for a double-reed pipe called an aulos sounds a G. The rhythm is complex and urgent, similar to that in modern Balkan folk music, and the microtonal slide is an ancestor, perhaps, of the mournful sounds (perhaps expressing Ξενιτιά or *xenitia*: a sense of catastrophic loss) recorded by Alexis Zoumbas in the *Lament from Epirus*.

The fragment is immediately recognisable as mournful because it shares features that we experience as sad in a lot of music we hear today. Most obviously, there are sigh-like passing notes which resemble a crying or sobbing human voice – the 'dying fall' prized by Duke Orsino in Shakespeare's *Twelfth Night* that is known in Western classical music as *appoggiatura* and in Klezmer as *krekhts*, or sobs. You can hear them in Tomaso Albinoni's Adagio in G minor (which narrowly trails Monty

Python's 'Always Look On the Bright Side Of Life' as a favourite tune at funerals in Britain), in 'Heaven Knows I'm Miserable Now' by The Smiths, and in 'Easy On Me' by Adele.

Then there is the drooping contour in the sequence of notes – a musical enactment of how grief and sadness tend to pull a body down and diminish its sense of energy. This feature is also widespread, and not just in Western music. *Sa-yalab*, a kind of performative or ceremonial musical weeping that is considered to be the most refined, reflective and beautiful among the Kaluli people of Papua New Guinea, follows a melodic contour of four descending tones. The first two are a major second apart, the next a minor third below, and the last a major second below, as in D-C-A-G. The pattern, reports the ethnomusicologist Steven Feld, is said by the Kaluli to derive from the call of a fruit dove.

A third feature of the music from the *Orestes* fragment that is also found in a lot of more recent sad music in the West is the minor interval, in which a sense of darkness or melancholy is often created by narrowing the space between two notes. (In *Orestes*, this is the minor third between B flat and G.) Many of the most moving works of lament in the European tradition are written entirely or almost entirely in a minor key, as well as deploying *appoggiatura* and descending note sequences. In Claudio Monteverdi's *Lamento della Ninfa*, for instance, which is in A minor, a single bass line on a descending fourth, A-G-F-E, underpins the dying fall in the soprano and accompanying instruments. In *Dido's Lament*, which is in G minor, Henry Purcell juxtaposes a five-bar ground-bass against a nine-bar sung melody with *appoggiaturas* and falling phrases in both the voice and strings. The opening chorale of the St John Passion by J. S. Bach, which is also in G minor, begins with bass instruments vibrating on the tonic while violins weave sixteenth notes around the other tones of the triad. Above this, a series of suspended minor dissonances are played piercingly by the oboes, first an E flat against another's held D, then F sharp against G, A flat against G, E flat against D and B flat against A natural. Tension and grief build to extraordinary intensity.

For all this, the association between minor scales and sadness is not fixed in the Western tradition. Think of 'Puttin' On the Ritz' by Irving Berlin, 'Golden Slumbers' by The Beatles and any number of other songs in which an interplay of both major and minor create a variety of moods. And beyond the Western tradition the sense of 'minor' as we hear it scarcely applies. Arabic music favours the *zalzal* – a quarter-tone halfway between what the Europeans call a major and minor third – to express a kind of sweet melancholy, and in the context of North Indian Raga, the flat third, which in the West we might hear as 'minor', evokes what the author Amit Chaudhuri calls 'reflectiveness'.

Other traditions have played formative roles in shaping much of what many who read this will experience as sad music today. Particularly influential are songs of call and response (perhaps shaped by Muslim sung prayers) that were brought from West Africa to the Americas by slaves. From the encounter in the United States of field calls, work songs and spirituals with European hymns and folk songs emerged gospel, jazz and the blues with its characteristic 'minor' tones such as the flattened fifth. It's from here that we have works such as John Coltrane's 'Alabama', a lament for four young girls killed in the bombing of a church by the Klan in 1963.

Recalling his childhood as a slave in Maryland in the 1820s, the abolitionist Frederick Douglass offers a striking first-hand account of what that root music was like. As slaves walked through the woods to the Great House to collect a monthly allowance, they would sing.

> Every tone was a testimony against slavery, and a prayer to God for deliverance from chains. The hearing of those wild notes always depressed my spirit, and filled me with ineffable sadness.

I have frequently found myself in tears while hearing them. The mere recurrence to those songs, even now, afflicts me; and while I am writing these lines, an expression of feeling has already found its way down my cheek . . .

According to Douglass, the songs – composed as the slaves walked along 'consulting neither time nor tune' – revealed both 'the highest joy and the deepest sadness'. The slaves would, he writes, 'sometimes sing the most pathetic sentiment in the most rapturous tone, and the most rapturous sentiment in the most pathetic tone'. There is a vital point here: song, whether 'sad' or not, and however 'simple', can express or carry more than one emotion – in this case, desolation at enslavement, joy at being alive in the woods as well as, for Douglass, a renewed fury and determination to free those in bondage.

Song is always part of something bigger, and even 'simple' music can affect a listener deeply. A good example is the oldest known complete written musical composition (as distinct from fragment) in the world. Inscribed in Greek 2,000 years ago on a tombstone near Ephesus in Anatolia are these lyrics:

Ὅσον ζῇς φαίνου	While you live, shine
μηδὲν ὅλως σὺ λυποῦ	have no grief at all
πρὸς ὀλίγον ἔστι τὸ ζῆν	life exists only for a short while
τὸ τέλος ὁ χρόνος ἀπαιτεῖ	and time demands his due

The notes and the rhythm to which they are set are given with marks and signs above each of the words. Recreating the tune is therefore quite straightforward, and what emerges is a light, almost danceable melody – a song you can learn quickly, and retain long afterwards. So while the tombstone is a marker of loss, it is a call to joy within the boundaries of time: to a kind of bright sadness. Something that Nick Cave says of Nirvana, Nina

Simone and other musicians he loves applies here: 'What we are actually listening to is human limitation and the audacity to transcend it.'

Music is capable of leaping the temporal gap between the moment in which it is composed and the moment in which it is heard. In a meditation on the nature of songs, the writer John Berger suggests that they have a dimension which is uniquely theirs. 'A song, while filling the present, hopes to reach a listening ear in some future somewhere.' Further, 'the tempo, the beat, the loops, the repetitions . . . construct a shelter from the flow of linear time: a shelter into which future, present and past can console, provoke, ironise, and inspire one another.' For the poet Anna Kamieńska, it is as if time is 'intensified, revitalised, recharged'.

Ted Hughes once said that writing is about trying to take fuller possession of the reality of your life. Songs can be a way of doing this too. Sad songs can help their creator, performer or listener work through and share difficult emotions in a relatively safe context. When Arlo Parks sings 'you're not alone,' she expresses for many of her generation an old truth: consolation is, as its etymology reveals, *with-solace* – the comfort in being together. And because songs and music more generally also have the capacity to express complex and changing emotions, they can open the singer or listener to the possibility of change.

In some situations, of course, sad songs can only reflect rather than alleviate suffering. In recent research into the perennial question of why so many people enjoy sad music, the psychologists David Huron and Jonna Vuoskoski found that those who do so score highly on measures of empathic concern (which they term 'compassion') and imaginative absorption ('fantasy'). When, however, levels of personal distress ('commiseration') are high, such enjoyment becomes all but impossible. 'Music, if anything, makes us raw, more susceptible to pain, nostalgia, and memory,' wrote the art critic Philip Kennicott in the wake of his mother's death, and experiences like his will be familiar to some who have lived through or observed bereavement or deep depression. 'If sadness is like a head cold,' says the writer and podcaster David Kallison, 'depression is like cancer.' Sometimes, for those feeling utterly wretched, no music can console. The pain is the only thing that's real.

After the atrocities of World War Two, which included the murder of both his parents, the poet Paul Celan wrote that only language remained reachable, close and secure amid all losses. And in his 1945 poem 'Deathfugue', it is as if music itself is consumed by death, when a camp commandant 'calls out play death more sweetly death is a master from Deutschland / he calls scrape those fiddles more darkly then as smoke you'll rise in the air'. But music can express defiance too, and it sometimes endures in the face of catastrophe. 'Joy is beyond sorrow, beauty is beyond horror,' wrote Olivier Messiaen, who composed his *Quartet for the End of Time* in a prisoner-of-war camp. And in an essay titled 'Humanity at Night', the philosopher Sarah Fine relates an account by Elie Wiesel of his encounter in the Auschwitz-Birkenau sub-camp of Gleiwitz with a violinist named Juliek, whom Wiesel remembered from Warsaw before the war. That night, 'in a dark barrack where the dead were piled on top of the living', Wiesel heard Juliek playing part of the Beethoven Violin Concerto. 'Never before,' he writes, 'had I heard such a beautiful sound . . . it was as if Juliek's soul had become his bow . . . [He] died that night.'

I sometimes think of this when I listen to a recording of Beethoven's Violin Concerto made by Jascha Heifetz with the Boston Symphony Orchestra under Charles Munch in 1959, just fourteen years after the end of the war. The sound of the violin can come so very close to the human voice, and Heifetz has a supreme ability to make it sing. I have always particularly loved the optimism and playfulness of the Rondo, and especially the sense of mystery in its second theme – a brief excursus from

D major to G minor played first by the solo violin and then in dialogue with bassoons and other instruments in the orchestra before it falls back into oblivion.

In the early 1980s the physician and essayist Lewis Thomas lamented that he could no longer listen to Mahler's Ninth Symphony with the mix of melancholy and high pleasure it had previously given him. Instead, he could only hear 'the door-smashing intrusion of a huge new thought: death everywhere, the dying of everything, the end of humanity'. The context was an especially tense period in the Cold War during which the United States and the Soviet Union were upgrading their thermonuclear weapons and were ready to launch on a hair trigger. Television presenters were talking about the possibility of limiting the numbers of murders in any given exchange to a few tens of millions. For Thomas, the cellos in what had been one of his most-loved pieces of music now sounded like 'the opening of all the hatches [of nuclear missiles] and the instant before ignition'. He writes that if he (around seventy at the time) were sixteen or seventeen years old he would want to give up listening and reading: 'I would begin thinking up new kinds of sounds, different from any music heard before, and I would be twisting and turning to rid myself of human language.'

We now know that a large-scale nuclear exchange was only avoided by sheer luck, including on one occasion quick thinking by a Soviet Air Defence Forces lieutenant colonel named Stanislav Petrov. The likelihood of this particular catastrophe has receded for now – or seemed to have done before the Russian invasion of Ukraine – but whether our civilisation (for all the benefits it delivers in the near term) is any less unstable is an open question. Climate change and the huge damage that human activities inflict on the non-human living world loom large. And there may be worse developments (alongside some that bring hope) around the corner. 'Perhaps we'll say nothing of earthly civilisation,' writes Czesław Miłosz, 'For nobody really knows what it was.'

I was not far off sixteen or seventeen when Lewis Thomas wrote down his late-night thoughts on listening to Mahler's Ninth. I have not thought up any new sounds, but I am always looking for them, as well as treasuring old ones. For me almost nothing surpasses the voice of the sea on a pebble beach, and there's a sense in which the Adagio of Mozart's Piano Concerto No. 23, in all its simplicity and understatement, says – in six or seven minutes, and without words – almost all that needs to be said about sadness, and beauty. While you live, shine.

Bashō

What exactly is the sound of a tear entering water? If you can imagine it, says the poet Alice Oswald, you are listening to the true note of the seventeenth-century poet Robert Herrick. Something similar could be said of Herrick's contemporary Matsuo Bashō, whose haiku – micro poems just seventeen syllables long – enact close observations of phenomena in the living world. And many of the finest of these haiku turn upon sounds, or their absence.

The sheer noise made by cicadas – typical of hot summer days in Japan – features in several haiku by Bashō. Scientists have recently discovered that, at 106 to 120 decibels, they are among the loudest of all insects, and Bashō marvels at the extraordinary life force expressed in their songs. In the stillness, he writes, it 'drills into rocks'. He also reflects with gentle wit on how ephemeral this is: cicadas do not live long, but by their cry 'you'd never know it'. Picking up the husk of a dead one, he marvels that it has 'sung itself utterly away'.

According to Philip Larkin, the most difficult kind of poem to write is 'the expression of a sharp, uncomplicated experience; the vivid emotion you can't wind yourself into slowly but have to take a single shot at to hit or miss'. This is what Bashō achieves again and again, but because haiku are so short the reader or listener can often only reflect fully on the nature or significance of their content in the silence that comes afterwards. An observation that Alice Oswald makes about poetry in general is especially apt in this case: 'the poem isn't always what happens in the words but is the trace that the words leave inside you as it vanishes'. In Bashō's 'sound haiku', the imagined sound continues to resonate after the poem ends. So, for example, midfield, attached to nothing, a skylark is singing. Or a rice-planting song in the back country becomes, simply, 'the beginning of art'. And coming across a village without bells, the poet asks, rhetorically, 'how do they live?' before panning out to 'spring dusk'.

Japanese tradition prizes *mono no aware*, or love for the beauty of transient things, together with *wabi*, or austere beauty, and *sabi* – a word which is hard to translate but whose meanings include 'ripe with experience and insight', and 'tranquillity, deep solitude'. Bashō matured in the context of this tradition, and his sound haiku express its essence. What voice, he asks, what song, oh spider, in the autumn wind? And in winter solitude – in a world of one colour – there is only 'the sound of wind'.

But Bashō also valued *karumi*, or a kind of joyful lightness. This is apparent in a haiku written when he was just twenty-three years old in which, during a spring breeze, laughter bursts out amid cherry blossoms. And the quality of *karumi* develops in the older Bashō to include gentleness and good-humoured irony towards the limits and illusions of the self. Even hearing the cuckoo's cry in Kyoto itself, he writes, 'I long for Kyoto'.

Throughout his life Bashō continued to hone his work towards *karumi*. In a talk with students towards its end, he said:

> now in my thoughts the form of poetry is as looking into a
> shallow stream over sand, with lightness both in the body of the
> verse as well as in the heart's connection.

Compare Henry David Thoreau a century and a half later:

time is but the stream I go a-fishing in. I drink at it; but while I
drink I see the sandy bottom and detect how shallow it is. Its thin
current slides away, but eternity remains.

With this lightness of touch, Bashō expresses, and offers, a
glimpse of a state in which we apprehend a larger reality. And in
this there is something akin to *enargeia* – the ancient Greek term
for an irruption of clarity towards 'bright unbearable reality'.
So a haiku by Bashō can move in the space between two short
lines from yellow rose petals to thunder . . . which turns out to
be a waterfall. The temple bell stops, he notes, 'but I still hear
the sound coming out of the flowers'.

Sound is at the centre of one of the best-known of Bashō's
haiku. This translation by Robert Hass is one of dozens:

> The old pond –
> a frog jumps in,
> sound of water

The poem could hardly be simpler and yet, centuries after
Bashō's death, it continues to resonate in surprising ways. Here
is one. Astronomers have suggested that the Sun, and hence the
Earth and other planets, may originate in an event in which
the Milky Way was – I like to think – something like the pond,
and a smaller passing galaxy was the frog. 'You have the Milky
Way in equilibrium, mostly calm,' explains Tomás Ruiz-Lara
of the Astrophysics Institute of the Canary Islands, 'and then
when [the small galaxy] Sagittarius passed it was like throwing
a stone in a lake. It created these ripples in the [density of the
Milky Way] so some areas became more dense and started
forming stars.'

A comparison between a seventeenth-century poem about
a frog and a pond on the one hand, and a recent finding
about star formation on the other, may seem far-fetched. I was

about to drop it when I found that at least one not totally crazy person had been thinking along similar lines. Here is *On Bashō's Frog*, an interpretation by the poet Michel Lara:

> pond stillness
> frog plunges in –
> universe ripples

Visible Sound

In 1912 the showman and self-styled 'nature singer' Charles Kellogg astonished the world with a new act. Kellogg was already famous in California and beyond for his uncannily accurate imitations of birdsong and his ability to summon wild bears, who would sit and listen quietly while he sang to them. He had counted the pioneering environmentalist John Muir among his friends, and drove around California in a truck made out of the trunk of a Giant Redwood tree mounted on the chassis of a Nash Quad. But here was something even more amazing: with nothing more than his voice, Kellogg could make a steady gas flame inside a long glass tube on the other side of the stage bob and dance, and by drawing the bow of a violin across a tuning fork he could put it out.

Extinguishing flames in this way became part of Kellogg's stage routine. Eventually, it caught the attention of scientists at General Electric, who in the 1920s set up an experiment that would either sniff out fraud or showcase the sound reproduction quality of their radios. While Kellogg sat in a radio studio in Oakland, a friend ignited a gas burner in front of a receiver forty miles away in San Jose and watched in astonishment as its flames, more than two feet high, were extinguished as soon as Kellogg began to broadcast. A few months later Kellogg repeated the experiment for the no less sceptical scientists at Berkeley. This time hundreds of listeners to the live broadcast wrote in to say that Kellogg had extinguished candles and matches that they had held up to their radios at home.

Remarkable as it was, Kellogg's act was just the latest in a long line of experiments in making sound waves and their impacts visible. In 1858 the physicist John LeConte had observed a flame pulse in synchrony to a musical beat, and flicker in response to trills on a cello. His report got the attention of John Tyndall who, alongside Hermann von Helmholtz and others, helped lay the foundation for our modern understanding of sound as, essentially, waves of pressure.

Even earlier, back around 1800, the physicist and musician Ernst Chladni had developed a remarkable means of making shapes with sound. In the technique he pioneered, a flat metal surface is mounted on a central stalk and sprinkled with a granular substance such as fine sand. When the bow of a violin is drawn along the edge of the plate patterns begin to form – hitched loops, the shapes of butterfly wings, radial arms like those of a brittle star or a heraldic sun in splendour. It looks like magic but there is a simple explanation: different regions of the surface are vibrating in opposite directions, and where they meet there are lines of no vibration. The sand jiggles away from the vibrating areas, which are called antinodes, and congregates along the lines, which are called nodes. The Chladni effect continues to fascinate people today. Millions have watched 'Cymatics', a video by the composer Nigel Stanford that shows water, oil and other substances forming strange shapes in response to electronic music and drumbeats. And there are practical applications: violin-makers have used Chladni figures to provide feedback as they shape the front and back plates of the instrument's body, with greater symmetry in the patterns meaning richer tones.

Sound can break things too. The famous trick in which a

soprano shatters a wine glass is no deception. Any particular glass shape has a natural resonance – a frequency at which it will chime if gently struck – and if the singer sings loudly enough at exactly this frequency the glass will resonate more and more, wobbling and warping until it shatters.

Extremely loud sound can do a lot of damage even without finding a critical resonance. Above 150 decibels, which is the intensity of a jet engine nearby, it can burst human ear drums, and above about 185 decibels it can kill by causing an air embolism that travels to the heart. Noise at this intensity has the potential to be used as a physical weapon, although so far militaries have usually preferred to degrade their targets psychologically by blasting them repeatedly with unwanted noise within the range that does not necessarily cause direct physical damage, including music the targets are likely to find obnoxious. Very loud sound can be one of the hardest things to bear when under bombardment. 'The noise,' a refugee from the shelling of the Ukrainian city of Kharkiv told Charlotte Marsden, a volunteer at the Polish border in March 2022, 'it just goes on and on, you can't think straight, the terror ...' Another refugee confirmed this: as well as physical damage, the Russian military sought to use what Nataliya Zubar, a resident of Kharkiv, described to the journalist David Patrikarakos as 'acoustic terror' – forcing people to flee due to the relentless thunder of their bombing. But noise doesn't have to be very loud to inflict psychological harm. The soundtracks of videos showing atrocities and trauma, warns Giancarlo Fiorella of the investigative journalism group Bellingcat, 'can leave as vivid an imprint on your mind as imagery'.

Charles Kellogg dreamed that his flame-extinguishing method might be used to fight fires, and showed it to fire departments in cities across the United States. He failed to convince any of them, but his dream did not die. In 2015 two engineering students named Seth Robertson and Viet Tran demonstrated an acoustic device that could reliably put out small fires with sounds in the bass range of thirty to sixty hertz. They suggested that there might be applications everywhere from kitchens to spacecraft.

But whether sound is used in future to hurt or to heal, there is another story about Kellogg that deserves attention. As related by the jazz critic and music historian Ted Gioia, it has the quality of a fable. Walking in New York City one day, Kellogg stopped suddenly at a noisy intersection and told his companion he could hear a cricket singing. The friend responded that this must surely be impossible given the din of traffic. But Kellogg looked around, crossed the street and pointed to a tiny cricket on a window ledge. The friend started to compliment Kellogg on his excellent sense of hearing, but instead of answering him Kellogg pulled a dime out of his pocket and dropped it on the sidewalk. The moment the coin hit the pavement it made a small pinging noise and everybody within fifty feet stopped and started looking for the coin. People listen for what's most important to them, Kellogg later explained.

Plato's Cave

In Plato's allegory of the cave, men see only the shadows of reality flickering on the wall of their prison. As a student in Cambridge in the Proterozoic eon otherwise known as the early 1980s, the Arts Cinema on Market Passage was my cave of choice. And the soundtracks – the echoes that accompanied the imagined realities of this magic lantern – were as important as the shadows themselves in shaping my dreams.

I remember being captivated by a scene in *Orphée*, Jean Cocteau's 1950 retelling of the myth of Orpheus, in which the hero becomes obsessed with broadcasts received over a car radio of a mysterious sequence of words and numbers. His companions are puzzled. 'Seems like nothing but meaningless words to me,' says one. No, insists Orpheus, 'The least of these phrases is much more than any of my poems . . . Where could they be coming from? . . . I feel certain they are addressed to me personally.'

Broadcasts like the one in this scene would have been familiar to many of Cocteau's contemporaries. It was, as he acknowledged, inspired by a number station – the means by which the BBC had broadcast coded messages to the French resistance during the occupation in World War Two. And this would not only have been the case in France. In *The Periodic*

Table, Primo Levi, who had fought with the partisans in Italy before he was sent to Auschwitz, recalls an 'intricate universe of mysterious messages, morse tickings, modulated hisses, deformed, mangled human voices which pronounced sentences in incomprehensible languages or in code ... messages of death ... the radiophonic Babel of war'.

Number stations, which broadcast sequences of digits that can only be decoded through a device known as a one-time pad, actually date to World War One. They have a clear purpose as secure channels for military communication and espionage, and they are still in use in some circumstances today. But to outsiders they can seem mysterious, and in *Orphée* the transmission is charged with a sense of metaphysical significance: a voice in the machine with a message we cannot quite grasp. And underpinning it there is also something of the essential marvel of radio itself. The power to transmit sound across previously insurmountable distances had so awed pioneers such as Thomas Edison and Guglielmo Marconi around 1900 that they had seriously wondered whether radio would make it possible to talk to the dead. Some claims made for the new technology in the 1920s were only slightly less fanciful. The Russian futurist Velimir Khlebnikov declared that the radio of the future would unite all mankind. More banally – and with hindsight, comically – signs in the studios for the first BBC radio broadcasts in the 1920s read: 'If you sneeze or rustle papers, you will DEAFEN THOUSANDS.'

The Cambridge Arts Cinema in the 1980s showed new films as well as classics. Godfrey Reggio's *Koyaanisqatsi*, released in 1982, broke new ground with what at the time was unprecedentedly high-quality time-lapse cinematography, showing ancient landscapes and the increasingly frenetic, fossil-fuel-based civilisation that was destroying them. For me and many others, though, it was the music by Philip Glass that held the sequences together. His score starts with a solemn passacaglia on the organ and a deep bass voice singing the word that is the title of the film and which means 'a life out of balance' in the language of the Hopi people. In the forty years since the film appeared, the destruction of non-human life on Earth has increased and emissions of greenhouse gases have greatly accelerated, turning dangerous climate change into a reality, so the message seems prescient. 'We can't say we didn't know,' wrote Bruno Latour in 2017. 'It's just that there are many ways of knowing and not knowing at the same time.'

But some of the most enduring sounds for me from that time feature in another film. Andrei Tarkovsky's *Stalker*, which had first been screened in the Soviet Union in 1979 and went on a limited release in the UK in the early 1980s, is a cryptic science-fiction story about a journey into a forbidden and dangerous 'Zone' where there is a room in which your deepest wish supposedly comes true.

Sound design in *Stalker* gives the film an unusual depth. Early on, a long tracking sequence follows the three protagonists as they ride a motorised trolley along an abandoned rail track into the Zone. The camera closes in on each of them in turn, rendering the landscape through which they pass a blur, and showing them mostly from the side or from behind so that their ears are centre screen. All three are watching and listening intently, but all we hear is the indifferent, steady clack, clang and hiss of the trolley over the rails. Gradually, the 'actual', or diegetic, sound that is part of the representation of the world on screen morphs into weird electronic noises that reach beyond it. We are, perhaps, hearing part of the sound of the characters' inner journey, and our own.

The scene represents 'the most straightforward journey imaginable,' observes Geoff Dyer in his book about the film, but it is also somehow 'full of all the promised wonder of cinema'. There comes to mind a phrase attributed to the surrealist poet

Paul Éluard: 'There is another world and it is this one.' Sound in cinema can expand both the inner and outer world, enriching the space of dreams and illusions.

Earworms

One of strangest and most insistent earworms I ever had – like musical hiccups in my head – began with a song called 'L'homme armé', or 'The Armed Man'. The piece is not widely known, but if you can imagine the most popular track on an album titled *Now That's What I Call Burgundian War Songs of 1453* then you have a rough idea. It was absolutely massive at the time, and it inspired works by a generation of musical superstars such as Guillaume Du Fay, Josquin des Prez, Cristóbal de Morales and Giovanni Pierluigi da Palestrina. It's a bold, lilting tune, and I could not get it out of my head.

But that was just the beginning. Some way into the first hour of my torment, I was startled to hear a piano playing Dmitri Shostakovich's Prelude and Fugue No. 1 in C major. This piece starts gently, almost like something for children inspired by Robert Schumann's *Kinderszenen*, but quickly shifts into more harmonically ambiguous and emotionally complex territory. A work less like 'L'homme armé' would be hard to find. And yet here were both going around in my head, sometimes consecutively, sometimes both at the same time.

In hour two I began to hear the opening riff of 'Day Tripper' by The Beatles. Eventually, and to my great relief, the Fab Four drowned out the late-medieval war song and Shostakovich, but by now I'd had more than enough, and I went for a long run in the rain. This seemed to do the trick. I have had other earworms since then, but not so far with the same intensity and persistence.

Research suggests that as many as 98 per cent of people get earworms at one time or another. They may be annoying but they are usually harmless, and they bear no relation to musical hallucinations in that those who hear them do not experience them as actually being 'out there' in the world, and they are not indicative of mental illness or brain damage. And, according to the neurologist and author Oliver Sacks, whose work I turned to for guidance after my extreme earworm day, they have nothing to do with the involuntary repetition of movements, sounds or words that can occur with Tourette's Syndrome or Obsessive Compulsive Disorder. Earworms, Sacks suggests in *Musicophilia*, exploit a natural human liking for repetition. 'Even as adults,' he writes, 'we want the stimulus and reward again and again.' In music we get it. 'Perhaps, therefore, we should not be too surprised . . . if [sometimes] the balance shifts too far and our musical sensitivity becomes a vulnerability.'

There is no surefire cure for earworms, but there is widespread agreement that exercise or engaging in a moderately difficult task can help. Go for a run like I did. Do a puzzle. Read a story. So when, one day, I started to hear Queen's 'Another One Bites the Dust' on repeat I dug out 'The Supremacy of Uruguay', a comic story by E. B. White published in 1933 (which may have been inspired by an earlier piece by Mark Twain). White describes how the military of that small South American country discover a powerful earworm in an American popular song and weaponise it. They mount phonographs on pilotless aircraft that play the earworm at high volume, send these out around the world, and quickly reduce the citizens of all other nations to gibbering wrecks.

At least things are not that bad, I told myself as, for the forty-seventh time, Freddie Mercury asked if I was hanging on the edge of my seat.

Noise Pollution

If you believe the headlines, the Covid-19 lockdown in Britain in the spring of 2020 was a good time to be a hedgehog. The animals reportedly took advantage of the relative peace and tranquillity to indulge in 'noisy lovemaking', and experts predicted a baby hedgehog boom. I don't actually recall hoglets, as baby hedgehogs are known, becoming a trip hazard later that year but it is true that, as human activity was tamped down, some of the sights and sounds of the non-human world became apparent in ways that many people had never known. As air pollution abated, millions of people in northern India saw the Himalayas on the horizon for the first time in their lives, and in thousands of cities around the world birdsong replaced traffic as the soundtrack of our days.

It was often asked that year if the birds were singing louder than before the lockdown, because it certainly seemed that way. In fact the opposite was true: birds that are common in urban environments, such as the white-crowned sparrow in North America, were actually singing about 30 per cent more quietly. They only seemed louder because other noise was down by roughly half. Researchers also found that within weeks of the start of the lockdown, birdsongs regained qualities that had last been recorded decades before, when cities were quieter. The white-crowned sparrows, for instance, extended their songs back down into lower frequencies which would normally be drowned out, and their songs became richer, fuller and more complex. It turned out that, in an instance of what is known as the Lombard effect, the birds had been 'shouting', just as people raise their voices on a construction site or at a noisy party.

Having to sing more loudly takes extra energy, and can stress birds out even to the point of causing more rapid ageing and death, so there were significant benefits to the birds from the reduction in ambient noise. In quieter conditions they could also hear their chicks, the sounds of predators and the warnings of other birds more easily. Competing males may have given each other more space and so have avoided fights.

This was not the first time that the adverse impacts of human-made noise, or anthropophony, on the non-human world has been demonstrated. In 2012, Jesse Barber and a team at Boise State University in Idaho showed that even what seem like modest changes in the nature and level of sound can have a surprisingly large impact. They built a half-kilometre-long 'phantom road' in an area where no road had ever existed, mounting loudspeakers on tree trunks along its route and playing recordings of traffic on a popular tourist route in the Glacier National Park in Alaska. When the speakers were switched on, the number of birds in the immediate vicinity declined by nearly a third, and several species fled the area entirely. But Barber and her colleagues found that some of the biggest impacts were on the birds that stayed. MacGillivray's warblers, for instance, stopped putting on the weight they needed to fuel their migration.

The impacts of noise pollution at sea are well demonstrated too. In 2001, researchers studying right whales in the Bay of Fundy between Nova Scotia and New Brunswick in Canada noticed a sudden drop in concentrations of metabolites that indicate stress in the whales' poop. The researchers had also been monitoring sound levels in the water, and realised that the fall coincided precisely with the sudden decline in human-generated noise as ocean shipping came to a halt after the 11 September attack on the Twin Towers in New York.

For humans, cranking up the volume, whether by revving a motorbike or playing music at dance-club volume, can be a form of self-expression and a way to have fun. Very loud sounds send vibrations through the body that, when willingly chosen, many people find pleasurable. Research at Manchester University suggests they may also stimulate parts of the inner ear that govern balance and spatial orientation, creating 'pleasurable sensations of self-motion'. They can be a way of blocking out the world – a rebellion or a catharsis, as it is with the cacophony of bus horns in the 2018 film *Never Look Away*. They can be an assertion of power, or a celebration of communal and religious identity – as at religious festivals in Mumbai.

But too much noise hurts humans too. The sound level in Mumbai, which is one of the world's loudest cities, can reach around 120 decibels – below the threshold for actual physical pain, but still enough to damage hearing in just a few hours. Chronic exposure to ambient sound at even a modest fifty-five decibels can be enough to delay reading and language development in children, interfere with sleep in both children and adults, and increase the risks of heart disease, strokes and other adverse effects on adult health. It is estimated that as many as 30 per cent of Europeans are exposed to excessive noise at this level from road traffic at night, and the World Health Organisation estimates that at least a million healthy life years are lost every year as a result. In the rapidly growing, high-density megacities of the Global South, noise levels are often even higher.

Cities have always been cacophonous, but sound levels increased dramatically with industrialisation in the eighteenth and nineteenth centuries. Contemporary observers were stunned by what they heard. 'Blast furnaces were roaring like the voice of the whirlwind all around,' wrote the historian Thomas Carlyle upon visiting an iron mill in Birmingham in 1824: 'the fiery metal was hissing thro' its moulds, or sparkling and spitting under hammers of monstrous size, which fell like so many little earthquakes . . . They were turning and boring cannon with a hideous shrieking noise such as the Earth could hardly parallel.' In 1845, the geologist Hugh Miller was almost overwhelmed as his train drew into Birmingham. 'In no town in the world are the mechanical arts more noisy,' he wrote; 'hammer rings incessantly on anvil; there is an unending clang of metal, an unceasing clank of engines; flame rustles, water hisses, steam roars, and from time to time hoarse and hollow over all, rises the thunder of the proofing house.' Even far away from centres of heavy industry the impact could be substantial and sometimes devastating. 'I have been assured that the effect of the white man is often felt up to 500 miles from their frontier,' wrote Alexis de Tocqueville in his account of the young and expanding United States in the 1830s. Often, notes the author Amitav Ghosh in a reflection on de Tocqueville's observations and their wider context, the noise made by colonists would drive away the animals on which Indians depended. And this noise pollution, argues Ghosh, was a factor in what he terms bio-political wars, in which the colonists degraded, fragmented and poisoned the living world for indigenous peoples.

Fast-forward a century or so, and motorised ground transport and aircraft added to the cacophony of twentieth-century cities. By the 1960s the impact on human health of noise, as distinct from those of air pollution, contaminated water or other side effects of industry and congestion, were increasingly well understood. In 1972, the United States Congress passed what was perhaps the first comprehensive legislative effort in the world to address the problem with the Noise Control Act. More legislation followed in the USA and elsewhere. Implementation was slow and sporadic at best, but there were some dramatic local examples of improvement. When, in 1975, a study found that children in classrooms in New York City Public School 98 facing a noisy rail track had fallen eleven months behind those on the quieter side of the building, measures were taken to quieten the track and soundproof the windows, and a follow-up study showed that the gap in test scores between classes on the two sides had disappeared. As a result New York extended its noise abatement measures, eventually installing rubber pads between rails and ties on tracks throughout the subway system. One of the leading experiments in noise abatement today may be in Paris, where restrictions on the use of motor vehicles could reduce anthropogenic sound levels to their lowest levels in decades.

The pain and distress caused by human noise, or what the ecological philosopher Ginny Battson calls *Anthrophonalgia*, sometimes seems to be almost inextricable from how the economy is organised, and in many places noise pollution shows little sign of diminishing. The worldwide quiet of lockdown in 2020 was temporary and, just as carbon emissions bounced back in 2021, so too did noise. In India, where the Ministry of Environment set in place nationwide noise pollution laws in 2000, and campaigners continue to battle for enforcement, many cities remain among the noisiest places on Earth. Elsewhere, noise pollution may reach parts of the world that have hitherto retained much of their non-human sonic richness. A new capital of Indonesia, for instance, is planned in East Kalimantan not far from the few surviving enclaves of ancient rainforest, which the indigenous Wehea Dayak people share with endangered orangutans, gibbons, hornbills and clouded leopards.

One of the most damaging forms of noise pollution on Earth results from the hunt for offshore oil and gas with arrays of air guns. These blast sounds, which reverberate through the water and deep into the rock beneath, can be as loud as 260 underwater decibels – six or seven orders of magnitude louder than the loudest ships. Deployed in large batteries, the air guns fire every few seconds as ships tow them up and down in surveys that can run for months over tens of thousands of square kilometres. In some years in the North Atlantic, dozens of surveys run at the same time, and a single hydrophone in mid-ocean can pick up their sounds off the coasts of Brazil, the United States, Canada, northern Europe and West Africa. Areas under intense survey can become all but intolerable for most marine life. Large animals such as whales and fish flee when they can, but animals at the base of the food web cannot, and may be devastated. In an experiment off the coast of Tasmania, a single air gun killed all the krill larva within more than a kilometre, as well as most other plankton. It is thought that the sound waves from the blast shook many of the animals to death, while those which survived the initial shock died soon after because they could no longer hear or feel the world around them.

The best way to prevent this kind of devastation would be to stop offshore exploration for fossil fuels – a policy that aligns with the urgent task of reducing the risks of dangerous climate change and ocean acidification. Noise pollution from ships, which are individually quieter than seismic surveys but collectively put more sound into the ocean, can also be greatly reduced. Improved design can cut sound by as much as 80 per cent. Unfortunately, this is likely to take decades to achieve across the global fleet. In the meantime, relatively simple measures can make a significant difference. A 20 per cent reduction in ship speed need not affect delivery times if planning is improved, but it can reduce carbon emissions by nearly a quarter and noise by up to fourth-fifths.

It is possible to achieve a quieter planet. If the world switched to renewable energy such as wind and solar combined with energy storage, global shipping would fall sharply because 40 per cent of all products transported on the high seas are coal, oil and gas. Unlike other forms of pollution such as plastic, which are likely to contaminate soils and oceans for a long time to come even if humans stop dumping them now, noise pollution will disappear as soon as we stop making it.

Britain, a birthplace of the industrial age, is one of the most densely human-inhabited places on Earth and among those most depleted of non-human life. Many of its soundscapes are among the most noisy and degraded. But even here we can get an inkling of what parts of a different future could be like. In her celebrated account of the rewilding on the Knepp estate in Sussex, Isabella Tree celebrates the return of turtle doves, which were once common in England but are now rare. As the land begins to knit itself back together, she rejoices at hearing once more – but also as if for the first time – their 'unmistakeable purring: soothing, inviting, softly melancholic'.

Joseph Monkhouse has gone a step further, piecing together electronically the soundscape of the Somerset Levels as it might have been more than two thousand years ago in the Iron Age. In his recreation of a time when human disturbance was minimal, Monkhouse brings together the calls of seventy-three different bird species that typically live in habitats of this kind – shallow open water, wet alder and willow woodland, reed swamp and sedge fen – to create a gentle but rich whole. 'The rewilding movement has opened my eyes to what Britain was like in the past, and what it could be again,' Monkhouse says. 'I find myself wandering around, looking at landscapes, imagining what it was like before and wondering how it might be one day again.' In a project titled 'Six Thousand Years of Forests' he recreates soundscapes of the past, starting in 3980 BCE and passing through the medieval period, and looks forward with two scenarios for how woodland in Britain might sound sixty years hence. In the first, the songs of the Earth are almost silent. Traffic and machinery predominate. In the second, the old tapestry of natural sounds is re-forming, allowing space for humans and non human life to thrive and dream in peace.

The Sounds of Climate Change

On 28 May 2008, a mass of ice about three miles across and a mile deep broke off the Ilulissat Glacier in western Greenland. Over the course of seventy-five minutes, huge chunks, many of them 1,000 metres or more from top to bottom, slid away and rolled over, thrusting their undersides hundreds of metres up out of the ocean as they did so. At one point, a dark shoulder of ice that resembled a whale of incredible size emerged groaning from the depths. Adam LeWinter and Jeff Orlowski, who captured the event on camera for the 2012 film *Chasing Ice*, suggest that 'the only way you can put it into scale with human reference is if you imagine Manhattan, and all of a sudden those buildings start to rumble and quake, and peel off and fall over and roll around . . . this whole massive city just breaking apart in front of your eyes'. Even through the distancing effect of a YouTube clip, the enormity of sights and sounds so far outside everyday human experience is compelling, and it is hard not to watch and listen again and again.

Calving, as the shearing of icebergs from a glacier is known, is part of a natural cycle, but it seldom happens on the scale witnessed that day at Ilulissat. And it is only one small part of a broad set of changes that, whether humans are watching and listening or not, are now taking place faster and at a larger scale than at any time in at least several million years. The melt rate of the world's glaciers caused by human emissions of greenhouse gases, already well above the background rate by the year 2000, has doubled in the last twenty years. Many of the glaciers that remain are likely to shrink and disappear in the coming decades.

The effects of climate change upon the glacial landscape often takes a quieter and more subtle form than the Ilulissat crash. When the writer Robert Macfarlane visited the Knud Rasmussen Glacier in Greenland, he heard a low rumble, rising in volume as he approached, of ice melt pouring into a moulin (that is, a deep shaft) which presaged an encounter with what he describes as 'the most beautiful and frightening space into which I have ever looked'. Matthew Burtner, composer of the album *Glacier Music*, describes glaciers as songful beings. 'They express their state through an intricate sonic outpouring which is the result of their melting [that generates] a rich complex of interwoven voices, threading together into a symphonic tapestry of noise.' When they reach the sea, glacier fragments continue to make noises: small chunks of floating ice are called growlers because they sometimes growl like animals as gas trapped inside escapes from them.

There can be a strange beauty in some of the changes taking place. Listening through special equipment to a melting iceberg off the Antarctic Peninsula, the journalist Jonathan Watts hears the sound of air bubbles escaping up through the deep interior and is 'transported – not, it seemed, below the ocean, but into a vast cavern, where it sounded as if water was cascading from a high ceiling, each drip echoing through the emptiness'. Permafrost can also make almost musical sounds as it thaws. 'It's like an orchestral piece,' says the geographer Julian Merton of a crater subsiding in the ground in Yakutia in the Russian Far East. 'In the summer, when the head wall is thawing quickly, you hear the constant trickle of water, like first violins. And then you have these massive chunks of permafrost, up to half a ton, that fall to the bottom with a big thud. That's the percussion.'

The impacts of climate change on life on Earth are also apparent in the changing sounds of forests and other ecosystems. For twenty years at the same time each year, the musician and acoustic ecologist Bernie Krause recorded the sounds of birds, mammals, amphibians and insects at Sugarloaf Park in California. When short clips from each year are played back to back, a dramatic diminution and fragmentation is clear. Elsewhere, researchers have found that a degraded landscape doesn't necessarily become more quiet. In places such as the Ecuadorian Amazon, disrupted ecosystems sometimes actually get louder within certain ranges of pitch, at least for a while, as incoming creatures compete and cross-talk to fill 'holes' in the soundscape.

Acoustic ecologists also monitor changes that take place outside the scope of human hearing. When insects and bats decline in numbers, whether as a result of large-scale pesticide use, changes caused by global heating or other factors, the ultrasonic soundscape empties out too. On tropical coral reefs, a 'dawn chorus' of fish and other animals falls silent as extreme heat events kill the majority of the corals on which they depend.

The phrase 'climate breakdown' has achieved some currency in recent years. It can give the impression of a world falling apart and ceasing to work. But this gives at best a partial sense of what is happening. Yes, rapid climate change threatens the viability of many species and ecosystems and, without quick action to reduce emissions, is likely to degrade or endanger much of human and non-human life. Barring a transition without precedent, man-made climate change could bring near-unliveable conditions to billions of people. But the climate itself is not 'breaking down'. Rather, as the late climate scientist Wally Broecker put it, 'the climate system is an angry beast and we are poking it with sticks'. With the net addition of energy in the form of heat it is, if anything, speeding up.

The sounds of climate change are not just those of diminution and disappearance, but also those of more powerful hurricanes, heavier rainfall events and more destructive floods, and larger and fiercer fires. They may prove to be the sounds of more human anguish and suffering as well because, other things being equal, a hotter world is likely to be a more violent one. Climate change may cause even more harm by increasing the likelihood of war than it does by increasing the likelihood of extreme weather.

'It's been a while since we have been able to turn to the natural world for reassurance, to map the arc of an individual life against the eternal cycle of the seasons,' observes the poet Kathleen Jamie; 'the feeling of being imperilled is now constant.' In these circumstances, one of the most precious sounds is your voice. 'The most important thing,' says the atmospheric scientist Katharine Hayhoe, 'is to talk about climate change.' Every fraction of a degree of global heating matters, every year matters, every action matters, and we need to discuss how we can improve our lives and those of others while lessening our adverse impacts and finding more ways to put our words into action.

Hell

Dispatching and butchering a human being, and then transporting the joints home for the pot is not hard to do once you have had some practice. Or so Anaru, a member of the West Miyanmin people, told the biologist Tim Flannery when he reminisced fondly about raids on the villages of other tribes in the highlands of Papua New Guinea in the mid twentieth century. The details are quite horrible, so if in doubt jump to the next paragraph. According to Anaru, you grab the victim from behind, and thrust a sharpened cassowary leg bone violently downwards into the gap between the collarbone and shoulder blade so as to pierce the lungs. Then you detach the head, arms and legs from the torso using bamboo knives. You gut the torso and tie it, much like a backpack, to the back of the person who is to carry it, and wrap the head carefully in a package of palm leaves so that it can be carried slung from a looped cane handle. A severed leg and an arm are thrown over each shoulder, to be grasped in pairs by the wrist and ankle, and the bearer begins the long trek home. The killer's village, which may have gone hungry for weeks or months before, will eat well for the next few days.

By 1984, when Anaru was reliving the good old days of his youth with Flannery, the West Miyanmin had all but abandoned the highlands and were camped around the Yapsiei station, a small central-government outpost near sea level to the north of their old homeland. Life here was much more peaceful. There was no cannibalism, but there were horrors of a new kind. The former highlanders had little resistance to malaria, elephantiasis and the skin diseases that thrive at lower altitudes, or to newly introduced ills like influenza. Almost all, Flannery writes, were affected by disease. Many of the women had grossly swollen breasts, while most of the men had grotesquely swollen scrotums or deformed legs. The mortality rate for newborns was 100 per cent, and the few surviving older children had the bloated stomachs that are a sign of malnutrition and chronic swelling of the spleen due to malaria. *Grile*, a form of ringworm which makes the skin flake in great concentric circles, was almost ubiquitous. The condition disfigures skin all over the body and produces a sweet, sick smell that permeates everything.

Flannery, who at the time was a young researcher on the trail of Goodfellow's tree kangaroo, the dingiso and other amazing creatures in the highlands of New Guinea, was full of respect and compassion for the West Miyanmin, but the Yapsiei station felt like hell on Earth, with sounds to match. Days of extreme heat and humidity were broken only by titanic thunderstorms. 'Occasionally these storms are so ferocious that they sound like the approach of jet aircraft screaming down the valley,' he wrote. 'When such storms hit, the place becomes chaos. Trees writhe in the first gusts of wind. Then, within seconds, nothing can be seen through the pelting rain – not even a hand in front of your face.' The noise was phenomenal: 'the thunder is so loud and continuous that it blocks out all other sound and it soon seems that the world has become strangely silent. The true silence that develops an hour or two later, as the storm makes its way downriver, is all the more eerie for this effect.'

There are countless hells on Earth. Societies with advanced technology have created some of the worst. The hell of industrialised war has seldom been more powerfully represented than by the artist Otto Dix in his images of the trenches of World War One, which portray cruelties inflicted on human bodies and whatever passes for their souls that equal anything imagined in the most terrifying medieval European art or Francisco Goya's *The Disasters of War*. Outside of war, a small but resonant example, for me at least, are experiments performed in the 1960s by the psychiatrist Robert Galbraith Heath, who reportedly 'wired-headed' two human subjects by implanting electrodes in the reward centres of their brains. The subjects could activate the electrodes by pressing a button, and reported feelings of extreme pleasure and an overwhelming compulsion to repeat. If at first this doesn't sound horrific, take a moment to think about it.

The hells of which humans dream are also many and various, and are not always sites of torment. Hades, the underworld of Greek myth, is, by Homer's account at least, a place of sad diminishment rather than punishment. As the shade of Achilles tells the still-living Odysseus, 'I would prefer to be a workman, hired by a poor man on a peasant farm, than rule as king of all the dead.' In the pre-Christian world of the Anglo-Saxons and Norse, *Hel* is a great hall where warriors who have fallen in battle dwell after death alongside a goddess who also bears the name. *Valhalla* means, simply, 'hall of the fallen'. *Hel* and hall actually derive from the same Proto-Indo-European word **kel*, which means 'to conceal, cover, protect'. For the Sumerians of the third millennium BC, all souls except those whose bodies have not received proper burial travel to a dry and dusty underworld known as *Kur*. Family members of the deceased pour libations through a clay pipe into the dead person's grave to relieve their thirst, and music can alleviate the bleak conditions for the most fortunate.

In other traditions, the sounds of hell are part of what makes it hellish. The sound worlds imagined by Dante Alighieri and

John Milton, two great poets of hell in the Christian tradition, are especially vivid. In the *Inferno*, the first thing that Dante describes after passing through the gates beyond which one must abandon all hope is the sheer din. 'Light is silent' but sighs and howls resound under a starless sky, bringing him to tears. There is a jumble of languages and deformities of speech, words which *are* pain, and an unending tumult that goes round and round like a sandstorm. The sheer intensity of the sound is physically shocking. In this, Dante was in accord with popular beliefs of his time: medieval dream-visions of hell often reported that horrendous noise was one of its most prominent features.

There is no music in Dante's hell, for that would be a form of blessing, grace or prayer. The closest we come is the prodigious fart of a fallen angel named Barbariccia which resembles a chorus of war drums, trumpets and bells. And the nearest to a musical instrument is the grossly bloated belly of a forger named Master Adam, which resembles a lute. For modern readers the latter may bring to mind the vision of hell in the right-hand panel of Bosch's triptych *The Garden of Earthly Delights*, which was painted around 1500 (and features on the cover of Deep Purple's first LP). There, one man is crucified on a lute and another impaled on the strings of a harp while a demon leads a choir singing from a musical score tattooed on a naked human bottom. (You can find online a rendition of the notes played by lute, harp and hurdy-gurdy as *Hieronymus Bosch Butt Music*; it is surprisingly gentle and tuneful.) Meanwhile, a woman wedged inside a giant hurdy-gurdy plays a triangle, and there's a huge pipe and a drum with someone trapped inside. On a disc on top of the head of a 'tree-man', demons lead souls endlessly around giant bagpipes which are reminiscent of a scrotum and penis. The devil's got a fiddle and, it seems, the devil's got a harp.

As for Milton's epic *Paradise Lost*, one of the first things that strikes the reader or listener is, as Philip Pullman observes, the sheer sound of the poetry itself. But, writing 350 years after Dante, Milton imagines a hell that is, first and last, characterised more by absence of noise than by mayhem. When Satan first speaks it is to break 'the horrid silence' and, demons apart, this is a hell of empty and echoing space. No sound, but rather silence audible. Moreover, such sounds as he and other devils make are far from horrible. There is, not least, a splendid defiance in Satan's words. When he says that it is 'better to reign in Hell, than serve in Heaven' the reader understands why William Blake quipped that Milton (who supported a revolutionary government that executed a king believed by most people at the time to have been ordained by God) was of the devil's party without knowing it.

Some of the fallen angels in Milton's hell try to fill the place with pleasant music. They make 'the warlike sound / Of trumpets loud and clarions ... Sonorous metal blowing martial sounds'. They march 'to the Dorian mood / Of flutes and soft recorders: such as raised / To height of noblest temper heroes old', and build Pandæmonium, their palace, to the accompaniment of 'sound / of dulcet symphonies and voices sweet'. It is when Satan leaves hell and passes through Chaos on his way to paradise to tempt Adam and Eve that he encounters cacophony – 'a universal hubbub wilde / Of stunning sounds, and voices all confus'd'. The devils who remain behind, meanwhile, 'sing / with notes angelical to many a harp ... Their song [is] partial, but the harmony ... suspend[s] hell, and [takes] with ravishment / The thronging audience.'

Beneath these sounds, however, Milton's hell is barren and dead, a place only of disembodied echoes. The contrast drawn with Eden could not be stronger. There, the air is rich with the sounds of life – 'of leaves and fuming rills', together with 'the shrill matin song / Of birds on every bough'. For Adam and Eve, the 'warbling' of birds and the 'murmuring sound / Of waters' have been part of their consciousness since their first moments.

They also hear the singing of angels who patrol every night. Heaven too is full of joyful noise. And, as the drama unfolds, the sounds that the devils make in hell steadily diminish until they are no more than a 'dismal universal hiss'. Hell becomes deathly quiet; its desolation reminiscent of the viewless winds blowing round about the pendant world of Mars as recorded by the Perseverance Rover in 2021.

I do not know much about heaven and hell, but I do sometimes feel that I have seen and heard both in my dreams. One hellish place (though it was not only that) was an enormous underground construction project on Mars to house the contents of the British Museum and other collections around a cavernous central hall. Great crowds were there, working maniacally. There was a hint of something like the horrifying duplicate existence of shadowy underground people in Jordan Peele's 2019 film *Us*, but my sense of dread inside the dream was relieved by a nice chat about human folly with the late Ursula Le Guin.

Heaven, or something like it, appears more rarely in my dreams. One time I floated effortlessly far above the Earth through space that was deep blue. It was a firmament by Giotto – gloriously rich and bright, fretted with golden stars which glowed softly like lamps above, below and on every side, and everywhere there was an inner music. I almost feel I can hear that music as I write but, like Dante in his account of heaven in the *Paradiso*, I cannot precisely recall it (*'canti / da mia memoria labili e caduci'*), and can only say that it surpassed anything that can be heard on Earth.

For eighty years a long spit of land on the coast of Suffolk named Orford Ness was the site for weapons trials by the British military. During the final forty years or so of that time, the delivery systems for successive generations of nuclear weapons were tested here. These included a gravity bomb called the WE177, which was configured to yield up to 400 kilotons, or about twenty times as much as the Hiroshima and Nagasaki bombs, and the Polaris missile, which could deliver up to 600 kilotons. Visiting in 2009, more than a decade after the last weaponeers had left, the ecologist Paul Evans said that the place 'seems to hold a vibration, a ghostly aural image of some terrible cataclysmic noise'. But today, after more than two decades of 'controlled ruination' in the hands of its new owners, The National Trust, non-human life is returning. The long shingle bank and the marshes it protects are home to many species of birds, and many more pass through on their migration. The Ness still carries the past on its slowly heaving back – 'It speaks bullet, it speaks ruin,' writes Robert Macfarlane, but it also 'speaks redshank [and] swift current'. A long stretch to the south of the former firing range was almost untouched by the weapons tests, and there, especially towards the tip, the Ness is pure form and movement. The shingle banks build and remake themselves, beyond human knowing, in sweeping shapes that, seen from above, look like curlicues and fiddleheads. Here is not silence but quiet – a quiet which, the poet Ilya Kaminsky suggests, frames like a doorway. If we walk through, what will we hear on the other side?

Healing with Music

'Shaman' – a term probably derived from the Tungusic word *šaman* for 'one who knows' – strictly applies only to the traditions of Siberia and Mongolia, but rightly or wrongly it has come to be used for traditional healers in many cultures across the world and, for all the many differences in these traditions, there are some striking similarities in practices and techniques, in which these healers use song, dance and rhythm as a gateway to the spirit world.

Typically, a shaman's apprenticeship involves learning various songs both from elders and from spirits they encounter in dreams. These songs are then used during the practice of healing, in which a shaman typically travels to the spirit world. Such may have been the case with a young woman who lived near what is now Bad Dürrenberg in Germany about 8,000 years ago, and who was buried with an elaborate headdress made from antlers and a necklace made from the bones of several wild animals, including a polished bone from the throat of a wild boar. Study of the base of her skull suggests she suffered from a rare condition that caused her to lose control of her body and enter trance states, and in such conditions the boar and other animals might have 'spoken' through her.

No less important to many shamans is the use of drums or

other forms of percussion to inspire an ecstatic state in which healing work can be done. This drum is more than just a musical instrument. In some instances it is referred to as a 'shaman's horse' because it enables a magical journey to the spirit world, and it is often decorated with images of birds, the moon, the sun, a rainbow or an arrow. The drum may be used to drive away evil spirits as well as to summon helpful ones, which are sometimes believed to be kept inside it, where they are at the shaman's disposal. The kemantans, or shamans, of the Riau in Indonesia believe that the sound of the drum is the voice of the spirit who inhabits it during the curing ritual. And this belief, suggests Ted Gioia in *Healing Songs*, is strikingly similar to that of Sumerians thousands of years ago. As such it is evidence for the antiquity of music in healing practices in cultures that are distant from each other. So too the myth of Orpheus, a musician of exceptional skill who enchants animals, humans and gods, and who undertakes a journey to the underworld to recover his beloved. Similar stories from Siberia, East Asia, Australia and Africa tell of the recovery of the living from the land of the dead.

Music has also long been allied to healing in non-shamanic contexts. The ancient Greek god of medicine Asclepius, whose serpent-entwined staff remains a symbol of medicine today, was the son of Apollo, the god of music, and the physician Hippocrates taught that healing the soul through music could heal the body too. This tradition continued in the Muslim world and was expressed through works such as the *Epistle on Music*, a tenth-century work by the Brethren of Purity in Basra. From at least the thirteenth century, Muslim hospitals, or bimaristans, contained music rooms for the benefit of patients, and from the seventeenth century, if not earlier, professional musicians performed regularly at the Al-Mansuri Hospital in Cairo and Nur al-Din in Damascus. Come healing of the spirit, come healing of the limb.

The Christian Church welcomed chanting in religious practice but was often suspicious of ecstatic rites and, outside of highly controlled contexts such as the military, the use of the drum. A fascination in Europe with harmony gave rise to elaborate attempts to apply it to the practice of healing. Drawing on the doctrine of the music of the spheres inherited from antiquity, some proposed that the harmonic intervals that governed the macrocosm could be applied to the microcosm of the human body in order to restore it to health. 'The medico-musicians,' noted the fifteenth-century scholar Ludovico Carboni, 'say that veins or arteries of the chest, move to a count of seven ... "on account of the consonance of the fourth".' Gioseffo Zarlino, a sixteenth-century music theorist, linked attributes of the soul to specific musical intervals. 'The intellectual part corresponds to the octave because it has seven intervals [corresponding to] mind, imagination, memory, cogitation, opinion, reason and knowledge. To the fifth, with its four intervals correspond the four divisions of the sense: vision, hearing, smell and taste (touch being common to all of them).' The fifteenth-century scholar Marsilio Ficino noted that certain diseases were 'said to be miraculously cured by certain harmonies,' and that the greatest musicians could 'mix [notes of varying pitch] in such proportion that from the many a single form arises which results not only in vocal power but also in heavenly power'. Such ideas live on to some extent today in claims that by singing or listening to 'solfeggio frequencies' one can keep the spirit, mind and body in a perfect harmony.

Fantastical ideas proliferated in early modern Europe. The natural philosopher, magician and playwright Giambattista della Porta, who flourished around 1600, wrote that flutes made of different materials could cure different diseases. Poplar, for instance, cured sciatica, while hellebore countered the effects of dropsy (oedema), and an instrument made of cinnamon wood remedied fainting. But more down-to-earth approaches also began to emerge. In the mid seventeenth century the polymath

Athanasius Kircher undertook to find a cure for an outbreak in southern Italy of tarantism – a form of what now looks like hysterical behaviour which was believed to result from the bite of a spider, and whose victims were encouraged to engage in frenzied dancing in an attempt to prevent death. In his *Musurgia Universalis*, a huge compendium of musical knowledge, and *Phonurgia Nova*, which explored acoustics and the influence of music on the human mind, Kircher did not turn to abstract ideas about harmony. Rather he took pains to gather evidence on the condition, and composed a series of what were intended to be curative scores. A 2001 recording by Christina Pluhar and L'Arpeggiata gives a sense of how. In pieces such as *Tarantella Napoletana, Tono Hypodorico*, a lively but disciplined rhythm accented with castanets, accompanies a simple harmony with variations in a way that might enchant and entrain the dancers so that they can work through their agitated condition towards a calmer state. We are in a sound world with one foot in the popular dance and the other moving towards the baroque of Arcangelo Corelli's variations on *La Folia* (a favourite of the late Umberto Eco, who used to play it on the clarinet with no discernible influence on the amount of folly in the world).

Not all attempts at musical therapy in early modern Europe were an unqualified success. In 1737 the castrato Farinelli, who had wowed audiences across the continent with his range, purity of tone and theatricality, was summoned to Madrid to sing for King Philip V, who suffered from depression and insomnia. Initially the king responded well, but it quickly became apparent that the revitalising effect of Farinelli's voice wore off almost as soon as he stopped singing, and the superstar ended up staying for a decade, during which time he sang the same few songs to his royal patient every night. After Philip's death, Farinelli continued to sing for the new king, Ferdinand VI, who spent his days wandering around the royal apartments banging his head against walls and refusing to be washed or shaved. This did not, apparently, cast doubt on the efficacy of the singing cure, and Farinelli retired a wealthy man.

Music therapy in the twenty-first century draws on a century of trial and error in clinical contexts in which it has been shown to help manage stress, alleviate pain, enhance memory, improve communication and physical rehabilitation and more. Listening to or making music is associated with relaxation in the autonomic nervous system. Heart rate and breathing slow down, and blood pressure, muscular tension and oxygen consumption reduce. There is a significant reduction in levels of cortisol, a marker of a stress response in the body, and an increase in salivary immunoglobulin A, a natural antibody. Listening to and taking part in music is associated with a release in the brain of neurotransmitters such as dopamine that are associated with motivation and pleasure, and the production of natural opiates in the body. In one study, spinal surgery patients who could control their own pain medication used about half as much when they had access to their favourite music.

Music and certain other sounds can promote well-being across an entire human lifetime. The human ear is almost fully developed by the fourth month of pregnancy and as a consequence babies become familiar with rhythms and melodies like their mother's voice, heartbeat and breathing patterns several months before they are born. Premature babies are thrust into a very different sound world, and this can cause them stress. A neonatal intensive care unit is mostly quiet, but intermittent beeps and buzzes from equipment and occasional sudden sounds at other times can startle premies, and it has been found that when they are surrounded with recordings of sounds they would have heard in the womb, such as their mother's heartbeat and voice, they show fewer signs of stress. Live music performed in the neonatal intensive care units has also been found to stabilise babies' heartbeats, reduce stress and foster sleep. With all babies born at term, lullabies improve mood and encourage sucking, and hence well-being.

Almost all young children who can hear enjoy music, but some on the autism spectrum can show particular sensitivity to musical sounds. For these children, music therapy can result in improvements in non-verbal and gestural communication, including eye-contact and turn-taking behaviours. Intensive music therapy can help children with attention deficit hyperactivity disorder, enabling them to tap into their emotional reactions in a safe and supportive environment and to learn how to recognise and respond to shifts in mood.

Even in deep unconscious states adults continue to hear the world around them. People in comas sometimes respond to gentle singing or the spoken voice of a loved one with signs of relaxation such as slower breathing and a change in brainwaves. Listening to music can be helpful for tackling depression associated with post-traumatic stress disorder, and for confusion and other challenges after a stroke. Playing music can also help people living with severe brain injury. Gabrielle Giffords, a US Congresswoman who survived a gunshot wound to the head, says that relearning to play the French horn was 'a huge part of [her] recovery'.

Music therapy can help with movement challenges such as those resulting from a stroke or from Parkinson's disease. In what is known as rhythmic auditory stimulation, a simple rhythm embedded in instrumental music can help guide patients to walk more steadily. It is thought that having an 'external timekeeper' helps the patient to synchronise their movements as opposed to relying on the internal timing signals from areas of the brain that have been damaged. In some instances, music can do more than this for patients with Parkinson's and other conditions. 'Those people who could not take a step could dance,' writes the physician Oliver Sacks of the therapeutic encounter, and 'some of those people who couldn't utter a syllable could sing ... while they sang they danced, they were able to move freely, almost as if their Parkinsonism and their other neurological problems were bypassed.' Patients with Alzheimer's can also exhibit dramatic changes. In a clip that circulated in 2020, a wheelchair-bound individual named Marta C. González listens to music from the climactic scene in *Swan Lake* and begins to move her arms and upper body to the music. A performance by a prima ballerina recorded in the 1960s is intercut and the moves seem to be a near-perfect match. It turns out that González is not actually the dancer shown in the archival footage, as the clip seems to imply, but the strength of her response to the music is not in doubt.

According to Sacks there is no significant carry-over effect with music in the case of Parkinson's: 'once the music stops, so too does the flow'. But there can be longer-term effects for people with dementia. 'Improvements of mood, behaviour, even cognitive function ... can persist for hours or days after they have been set off by the music.' He reports the case of a patient with Alzheimer's named Woody for whom remembering anew that he can sing was profoundly reassuring. 'It can stimulate his feelings, his imagination, his sense of humour and creativity, and his sense of identity as nothing else can ... It can give him back to himself and, not least, it can charm others, arouse their amazement and admiration – reactions more and more necessary to someone who, in his ludic moments, is painfully aware of his tragic disease and sometimes says that he feels broken inside.'

The number of evidence-based interventions with music therapy continue to grow. I recently learned, for example, that taking up the didgeridoo can reduce problems caused by obstructive sleep apnea. I suggest as a slogan: 'With Didgeridoo, Snores Are Few.' You're welcome. But practitioners remind us not to oversimplify. The psychologist John Sloboda stresses that music is not like a vitamin provided to 'consumers' or 'patients', whereby a given piece of music has a given effect. No single piece of music can be prescribed as a universal stress buster or relaxation aid. Rather, the larger context is crucial. A vital dimension for music therapy, writes the psychologist Victoria Williamson, is 'the less tangible but very real [influence] of human contact, communication and empathy, guided reflection and emotional support'. As Ted Gioia puts it, 'there is no magic chord progression, no secret drumbeat, that will unlock the body's mysteries. Instead, the music works by reaching out, embracing larger and larger wholes.'

Making music oneself can be intimidating and even seem redundant for many of us in a world where the sounds of great artists are a click away. But the benefits to mental and physical health of active participation are well demonstrated. Singing in a choir is a good example. An especially dramatic demonstration of this was recorded by the otolaryngologist Alfred Tomatis at a Benedictine monastery in France in the mid 1960s when the monks curtailed their practice of daily chanting as part of a modernisation drive. Within a short span of time they became listless and exhausted, easily irritated and susceptible to disease. When, a few months later, singing was reintroduced, the monks were soon firing on all cylinders again. And one doesn't even have to be a good singer, let alone a monk, to see benefits. Simply by singing together with others one can co-create something bigger and different. 'The sharps and flats tones of amateur voices combine into a perfection that few of the singers could have obtained on their own,' writes Oliver Burkeman in his book on how to spend the 4,000 weeks of a typical lifetime. From my own experience of a community choir I'm not sure about the word 'perfection'. There is improvement over time, but more important is that the singing helps to bring us together, and support each other. A recent large-scale study in England showed that singing helps new mothers experiencing postnatal depression, and the

NHS is looking to increasingly roll out music therapy as part of 'social prescribing' for a variety of conditions.

Many of us in the largely urban societies of the West may have a lot to learn, not just from monks and community choirs but also from other traditions such as Native American healing rituals, which embrace the physical, spiritual and emotional health of patient, family and community. Some of these rituals are now a thing of the past, but others endure and evolve. One such is the Ojibwe practice of Big Drum – a gathering to mourn, process grief, understand loss and connect people to place. 'There's two words we use a lot at drum,' Joe Nayquonabe tells the writer David Treuer, 'wiidookodaa- didaa and zhawendidaa. Let's help each other, and let's care for each other.'

There is also a compelling case for extending what we think of as our community. For the !Kung of the Kalahari, writes the anthropologist Richard Katz, healing is 'more than curing, more than the application of medicine. Healing seeks to estab- lish health and growth on physical, psychological, social, and spiritual levels; it involves work on the individual, the group, and the surrounding environment and cosmos.'

There is a place for music as a sedative. If you're looking for a potent one try *Weightless Part 1* by Marconi Union – a slow- pulse ambient track which is considered so powerful that it's recommended not to play it while driving. But in a world out of balance there is also a need for human music and sounds of the natural world that wake us up and help us pay atten- tion. The sixth assessment report of the Intergovernmental Panel on Climate Change published in July 2021 warned that dangerous changes to the global climate may be inevitable and irreversible without unprecedentedly rapid reductions in emissions of greenhouse gases. 'It's very easy to feel crushed by a barrage of unbearable realities,' wrote Joëlle Gergis, a lead author of the report. 'But . . . I want to say that humans have the inherent goodness to turn this around.' If she is right, then sweet sounds that bring delight may help us find strength and joy for the long and difficult path ahead.

Healing with Sound

In 'A Very Old Man with Enormous Wings', a short story by Gabriel García Márquez, a bedraggled angel who has fallen to Earth in a small town in Colombia briefly becomes a minor local attraction. Some of those who come to see him are invalids who seek relief from troublesome noises. There is a poor woman who has been counting her heartbeats since childhood and has run out of numbers. There is a man who can't sleep because the sound of the stars disturbs him. Odd noises afflict the angel too. A doctor who listens to his heart hears so much whistling that it seems impossible to him for the angel to be alive.

The doctor in Márquez's fable is doing something that members of his profession have been doing for a long time. Auscultation, or listening to the internal sounds of the body, has been a part of diagnosis since at least 300 BCE when Herophilus of Chalcedon, who was convinced that the human frame made its own hidden music, started listening to his patients' heartbeats. In the early nineteenth century, the physician René-Théophile-Hyacinthe Laënnec found he could hear the sounds in his patients' chests clearly through a rolled piece of paper and avoid thereby the need to put his ear against bodies, which were often unwashed. A keen amateur musician who liked to make flutes, he fashioned a wooden tube to make the first stethoscope (from Greek *stēthos*, 'breast', and *skopein* ,'look at'). Two hundred years later, listening (auscultation) with a stethoscope is still often one of the first steps in a physical examination, along with looking (inspection) and feeling (palpation). 'The stethoscope is still by far the quickest way of hearing what's going on in a patient's chest,' the doctor and writer Gavin Francis told me. 'In my work as a physician it is particularly useful when checking newborn babies, whose hearts do these extraordinary contortions in the first few weeks after birth. The holes in the heart that are needed for life in the womb gradually close over, and if this doesn't happen you get all sorts of sounds made by blood flowing with turbulence, or even in the wrong direction.' After talking with Francis, I wondered if Márquez's angel may have been suffering from pericarditis, a condition in which the protective sac surrounding the heart roughens on the inside, and each heartbeat heard through a stethoscope is accompanied by a rustling sound known as a pericardial rub.

Francis also explained to me how auscultation can help in the diagnosis of lung and bowel conditions. In what is known as whispering pectoriloquy, the doctor asks the patient to whisper while they listen to the chest cavity. The sound is much louder than usual when there is increased density within the lungs, but is muffled when there is fluid around the outside of the lungs – a condition known as pleural effusion. In another technique, the patient is asked to say a word or phrase with lots of 'n' sounds in it. If there's consolidation in the lungs you hear the 'n's much more clearly and loudly. This is known as vocal resonance or aegophony, from the supposed resemblance of the sound to the bleating of a goat (Greek αἴξ: *aig-*). In the case of the bowels, a doctor will listen out for borborygmus: the rumbling and gurgling sound of healthy digestion. In an obstructed gut the noise tends to be much higher-pitched, 'tinkling', and almost constant as liquid tries to squeeze through twists in the bowel. Total silence is a bad sign, indicating digestion has shut down.

Medical students also learn how vibrations made by gentle tapping, or percussion, of body parts can reveal the presence of excess fluid, a tumour or something else that needs attention. This sounds like a difficult technique to master. 'I had a tutor who told us to put a two-pence coin under a phone book and try to tap out where it was,' said Francis.

An examination with a stethoscope or by percussion can reveal many health concerns, but medical ultrasound, in which very high-frequency sound waves are bounced off parts of the body's interior and collected to form a picture, can provide a level of detail that neither can begin to match. An echocardiogram – a picture of the heart made with ultrasound – shows the condition of its valves and muscles and how they are functioning with precision. Ultrasound can reveal patterns of blood flow, and vital clues about the condition of organs such as the lungs, liver, kidneys and even the eyes. It can also conjure an image of an unborn child, as the waves bounce harmlessly back from solid tissue and bone in the foetus. It is, truly, an astonishing thing that our first glimpse of the greatest wonder and trial of our lives, parenthood, comes in the form of a fuzzy black and white smudge made from sound.

Ultrasound is used in treatment as well as examination. At high intensity it can break down kidney stones and gallstones into pieces small enough for the body to expel more easily. In phacoemulsification, a form of cataract surgery, ultrasound waves can be focused at high energy to break down a clouded lens, which is then removed. (The technique was invented by the surgeon Charles Kelman, who also played jazz with Dizzy Gillespie and Lionel Hampton.) Ultrasound can ablate tumours and other tissues, and facilitate the better absorption of various drugs as part of chemotherapy and other procedures. Many new therapies are conceivable, but some avenues of exploration may prove more promising than others. A team in Arizona is looking at whether low-intensity ultrasound can be used to quieten parts of the brain that are active during episodes of negative, self-centred rumination and distress that are sometimes known colloquially as 'washing-machine head'. Given how little is known about many aspects of brain function, and how poorly understood the mechanisms of mood and anxiety are, the approach looks like a long shot at best. Nevertheless, Shinzen Young, a Buddhist monk associated with the research, remains hopeful. 'The technology scares me,' he says, '[but] the future without a radical improvement scares me a lot more.'

A lot of research indicates that natural sounds within the normal range of human hearing – in particular vibrant bird-song – can decrease stress, with significant psychological and health benefits. For the writer Lucy Jones, the noises of the natural world, alongside its sights, smells and tactile qualities, played an important role in her struggle to overcome addiction, come to terms with a sense of ecological loss, and find a more balanced way to live. 'Nature softened the voices in my head and stabilised my mood,' she writes.

Medical practitioners and others are looking at ways to adjust the soundscape in hospitals to further patient well-being. 'Too much unwelcome noise can be disruptive, but silence can be disconcerting as well,' says Victoria Bates, who leads the project 'Sensing Spaces of Healthcare: Rethinking the NHS Hospital'. It is, she suggests, a question of striking a balance. 'Rather than trying only to eliminate noise, we might also try to listen to it more closely: what sounds are perceived as noise and why? If the sounds of hospitals are heard not as a cacophony but as a mingling of purposeful and important sounds, could they be less distressing?'

Musicians and artists are also exploring these questions. In 2013 Brian Eno trialled *77 Million Paintings*, an 'ever-changing healing soundscape' for the reception area of the Montefiore Hospital in Hove in England. The work was politely received, but it is clear that no one set of man-made sounds suits everybody. 'One person's "soothing" is another person's "noise",' observes Bates. The sound artist Yuri Suzuki reckons that any new sonic design appeals to about 40 per cent of any given client group at most. Noting this, the artist Sally O'Reilly suggests, half playfully, that hospital sounds be tailored for each patient. 'We will explore the possibilities of scraping home environments for calming, familiar sounds of strip lights, vacuum cleaners, washing machines, traffic,' she writes. 'We will build up a library of ambient samples, from which the patient selects a personal mix.'

For the common spaces of a hospital – and for any place where we may pass from one future to another – perhaps we could move beyond a soundtrack of man-made noises and in their place introduce soft sounds of the natural world. 'No air is sweet that is silent,' wrote John Ruskin; 'it is only sweet when full of low currents of under sound – triplets of birds, and murmur and chirp of insects.'

Bells

A peal of English church bells is part of my early childhood. On summer evenings when it was still light at bedtime I would hear the sound coming across the London rooftops and through the open windows of my bedroom on the top floor of our house. On weekend visits to my grandparents in Hampshire the bells would echo through the hangers – steep wooded hillsides – around the village. That cascade of noise, bouncing off the curves of the land, and sometimes reverberating from more than one place at almost the same time, seemed to both describe and express the place, and I came to associate it with openness and joy – a sense of a presence deeply interfused. I still sometimes experience this today – marvelling that sound, which is normally an elusive and rapidly passing phenomena, seems to stand high above in the air, like a perfume or a mist illuminated with golden light. It is, as Henry David Thoreau observes, as if the vibration were a property of the air through which it passes. I also love how the sound can both make one's whole body vibrate and connect the listener to the greater world. 'If we opened people up, we would find landscapes,' says the film director Agnès Varda. 'I had been my whole life a bell, and never knew it until at that moment I was lifted and struck,' writes Annie

Dillard. Bells make sounds on the edge of music, and can alter our sense of space and time.

Our chimpanzee cousins thump the ground with sticks and fists to express emotion. Gorillas beat their chests. And humans have probably been finding other objects to thwack in order to make impressive sounds for at least as long as we have been human. Over time these other things have generally been one of two main types. The first, known as membranophones, are instruments in which the sound is made primarily by striking a stretched membrane such as an animal skin over a hollow drum. The second, idiophones, make a sound when the main body of the instrument is struck, and in this category ringing rocks, rock gongs and sounding stones date back at least tens of thousands of years. The elaborately carved megaliths at Göbekli Tepe in Anatolia, which were erected between about 9500 and 8000 BCE, are sounding stones of this type, and when struck they make very low sounds that can be felt in the body like the rumble of an underground train.

Bells are idiophones that have been made possible by one of the most momentous developments in human history: metallurgy. The power to create materials stronger and more resonant than stone and bone must have seemed like a kind of magic to those who first mastered it. Something of how they felt may be glimpsed in stories of great antiquity such as 'The Smith and the Devil', in which a cunning human tricks a supernatural being into giving him metalworking power, and even in the legend of Arthur and the sword in the stone. Gold and copper, marvellous in their sheen and allure, were smelted first, from around 7000 BCE, but a breakthrough came around 5000 BCE with bronze – an alloy of copper and tin which is harder than either. And it is from bronze that bells, along with weapons and countless tools, were first made. They still are. When we hear bells we hear the sound of the Bronze Age.

The oldest surviving metal bells are from Shang-dynasty

China (*c*.1600–1064 BCE). At first they were small and were probably made to hang around the necks of animals. But over time the Shang produced ever more spectacular bronze ritual objects: vessels, weapons and musical instruments including cymbals and larger bells. During the Zhou dynasty (*c*.1046–256 BCE), sets of tuned bells called *Bianzhong* were among the prize possessions of high-status individuals. When the Marquis Yi of Zeng died in 433 BCE he took with him the bodies of twenty-one young women, an armoury of bronze weapons, elaborate vessels and fittings for chariots, and a *Bianzhong* consisting of sixty-five bells. Some years after archaeologists excavated his tomb in 1978, it was found that the bells could still be played. Each bell, or *zhong*, is oval rather than circular in cross-section and makes two distinct notes, typically a major or minor third apart, depending on where it is struck. The whole set can reproduce twelve tones per octave over the range of a human voice, and it is thought the bells were used to play scales with six tones. We may imagine something halfway between the *slendro* (five-note) and *pelog* (seven-note) scales played on *bonangs*, the kettle-shaped gongs used in Indonesian gamelans today.

Large bronze bells were also cast in antiquity in South Asia. From early on these appear to have had circular mouths much like those commonly found in India and Europe today, and were used in rituals. *Ghanta*, as they are called in Sanskrit, are charged with meaning in Hindu, Jain and Buddhist practices that have endured for millennia. In Hinduism, the curved body of the bell is said to represent *Ananta*, or infinity, while the clapper represents *Saraswati*, the goddess of wisdom and knowledge, and the handle *Prana Shakti*, the vital force. The sound issuing from the bell is sometimes said to be the sacred *Aum*, signifying the ultimate reality, *ātman*. Hindu devotees will often ring the bell that hangs by the entrance to the inner sanctum of the temple to inform the deity of their arrival, dispel evil, disengage the mind from troublesome thoughts and focus on the divine.

Round bells like those of South Asia were probably introduced to China along with Buddhism towards the end of the Han dynasty in 220 CE or in the turbulent period that followed. By the time of the T'ang dynasty, which began in 618 CE, they were widely used in Buddhist temples across China and the older-style oval-shaped bells fell out of favour. The technology of casting large bells also reached Korea and Japan, along with Buddhism, at around this time. Buddhists in many places also favoured standing bells and singing bowls, which are in essence inverted bells with the rim uppermost. With their especially pure tone, these continue to be widely used. 'What an X-ray is to a conventional photograph the singing bowl is to . . . the Western orchestra,' comments the music writer Ted Gioia. 'No musical instrument is more penetrating.'

In Japan, *bonshō*, as the large bells are known, have played a major role since the seventh century in Buddhist ceremonies including New Year and the Bon festival, when ancestors are honoured. The oldest surviving example, and probably the oldest bell in the world still in use, was cast in 698. The largest, which was commissioned in 1633, weighs seventy-four tonnes – about as much as seven full-grown African elephants – and needs up to twenty-five people to sound it. The low tone and deep resonance of a large *bonshō* mean it can be heard up to twenty miles away. This led to their use as timekeepers, signals and alarms.

Bonshō are usually struck with a suspended beam rather than with an internal clapper. The sound of the bell is made up of three parts. The impact of the strike, called the *atari*, is ideally a clean, clear tone. This is immediately followed by the *oshi*, a prolonged reverberation for up to ten or so seconds. Finally comes the *okuri*, the resonance that is heard as the vibration of the bell dies away, which can last up to a minute. We may be reminded that while English has a single word for 'time', Japanese has many. Some derive from the ancient literature of China, or from Sanskrit, notes Anna Sherman in *The Bells of Old Tokyo*, 'Japanese has borrowed a vocabulary for vastness, for the eons that stretch out past imagination towards eternity: *kō*. Sanskrit has also lent a word for time's finest fraction, the *setsuna*: "particle of an instant".'

The Japanese celebrate *bonshō* in stories both real and imaginary. In the medieval epic *The Tale of the Heike* it is said that the sound of the Gion Shoja temple bells 'echoes the impermanence of all things ... The proud do not endure, like a passing dream on a night in spring; the mighty fall at last, to be no more than dust before the wind.' And in folk tales, the monk, ascetic, warrior and all-round superhero Benkei is said to have dragged the three-tonne bell of Mii-dera up Mount Hiei all by himself. Other legends suggest that *bonshō* can be heard in the underworld.

Since World War Two, bells in Japan have frequently been associated with prayers for peace. Visitors to the memorial to commemorate the victims of the atomic bombing of Hiroshima are encouraged to ring one of its three bells. The surface of one shows a map of the world, the spot where it is struck shows an atomic symbol, and the bell is inscribed 'know yourself' in Greek, Japanese and Sanskrit. There is also a Japanese Peace Bell, installed in 1954, outside the United Nations Headquarters in New York City. In 2017 the Ukrainian city of Mariupol installed a Peace Bell in a public park to celebrate safety and well-being. The city was almost completely destroyed during the Russian invasion of 2022.

In Europe, small rattle-like bells of various shapes called *tintinnabula* were common in the Roman period if not long before, but the story goes that large bells with circular mouths like those found in churches today originate in a dream of Paulinus, a fifth-century poet, senator and bishop who fell asleep in a field of flowers and was woken by angels playing upon them.

Paulinus later became a saint, and a festival of lilies in his honour still takes place in Nola in southern Italy, where he was bishop, every midsummer. Grimly, and a bit weirdly, the patron saint of bell founders is Agatha of Sicily, supposedly because the shape of her severed breasts resembles a bell.

Legends aside, there is evidence for the casting of round-mouthed bells in Europe from around Paulinus's time. Whether these were derived from or inspired by South Asian examples is not known, though it seems likely. Round handbells, known as *nola*, and bigger bells for monasteries and church towers called *campana* gradually spread across western Europe from about 530 CE as the Benedictines established bell foundries in their spreading network of monasteries. The Anglo-Saxon historian Bede describes bells of this kind in England in the eighth century. The oldest surviving church bell in Europe is thought to be one donated by an abbot named Sansón to a monastery in the mountains outside Córdoba in the year 930.

The development of artillery in Europe in the fourteenth century boosted bell production, because cannon used almost exactly the same alloy and were cast by similar methods, often in the same foundries. Big bells were a sign that you could make big guns, and by the fifteenth century super-size bells were being created for super-size buildings. The Bell of the Three Kings installed in Cologne Cathedral in 1437 weighs nearly four tonnes, and the *Pretiosa*, the larger of two bells installed in 1448, weighs 10.5, or as much as one large elephant.

As in Japan, bells in Europe were put to various uses. Smaller, hand-rung bells were common on ships, where they could mark the change of watch, raise an alarm, indicate position or toll a death. Think of Shakespeare's *The Tempest* (c.1610): 'Sea-nymphs hourly ring his knell: Ding-dong / Hark! now I hear them – Ding-dong, bell.' In England, a rural parish typically extended as far as its church bell could be heard: the boundaries were a kind of map made of sound. As well as drawing people to service, bells were used for warnings, celebrations and funerals. In monasteries they would mark the time for prayer. Chiming clocks that rang bells or gongs automatically had first been tried in China as early as the eighth century, and a striking clock on the Umayyad Mosque in Damascus in Syria was described in 1203. In Europe, mechanical clocks that struck bells began to spread in the 1300s, first to the richest cathedrals and then more widely.

Improvements in casting techniques in the sixteenth century led to new uses, including in music. Founders in the Low Countries developed the carillon: a set of tuned bells that could be played with levers or keyboards. The instrument was perfected in the early seventeenth century by the brothers Pieter and François Hemony, who worked with Jacob van Eyck, a blind musician and nobleman. In 1633 Van Eyck described the main tones and overtones that a bell makes in terms of the Western harmonic system. The loudest, called the prime or strike note in English, mainly resonates with four others: the nominal, which is an octave above the prime; the hum, an octave below; and the tierce and quint, which are a minor third and a fifth above respectively. Van Eyck persuaded the Hemony brothers to turn their bells on lathes and to shave and chisel down the interiors so that the prime and principal overtones of each bell stood out more clearly in the way we are now accustomed to hear in most western European bells.

Early in the seventeenth century, tuned bells were adopted in England for what is known as full-circle ringing. In this practice, the bells – of which there are typically six to eight weighing anything from around fifty kilograms (about 110 pounds) to more than a tonne – are hung from bearings and swung through more than 360 degrees by a rope that wraps round a circular wheel first one way and then the other. As the mouth of each bell swings downwards, the clapper strikes the part of the bell around the mouth, called the soundbow, to make it chime. As it swings up again on the other side of the arc, the clapper comes to rest against the soundbow and dampens its vibration so that the sound decays away quickly. This effect – almost the opposite of the long resonance of a Japanese *bonshō* – means that when bells in the set (or 'ring of bells') are sounded in quick succession their individual strikes do not merge altogether into a blur but remain partly distinct. And because an experienced bell-ringer can change the time between each of their strikes by using the rope to control the next swing of their bell, the order in which the bells sound can be changed with every cycle through the set. What starts with a simple sequence of descending tones can branch into any one of thousands of variations: ringing the changes. A member of the Oxford Society of Change Ringers, who set up shop in 1734 and have been practising every week ever since, told me that they can consistently keep each ring to within two-hundredths of a second (twenty milliseconds) of its place within a sequence in which each toll is a fifth of a second (200 milliseconds) apart.

In Russia, bells developed along a different trajectory from western Europe. The Orthodox Church came to regard them as entities that expressed sanctity in sound as icons do in paint. With a prohibition on musical instruments in religious service, bells were not tuned to make particular harmonics stand out as they were in the west, but rather to produce the fullest possible range of tones across the spectrum. 'A Russian bell must sound rich, deep, sonorous, and clear, for how can the voice of God be otherwise?' a priest involved in the revival of traditional practices told the author Elif Batuman in 2009. 'It must be loud, because God is omnipotent. Above all, Russian bells must never be tuned . . . The voice of a bell is understood as just that. Not a note, not a chord, but a *voice*.'

For hundreds of years bell towers were often the highest buildings in Russian towns and cities. Bells were the 'voices of the Russian sky' calling the faithful to prayer. Many were given names, and some were seen as magical beings, able to prevent the spread of diseases. The greatest bells were honoured objects of state. Ivan the Terrible would, supposedly, climb up the belfry at Alexandrovskaya Sloboda, which is nearly fifty metres (164 feet) high, every day at dawn to ring the matins himself. When one of his sons died in the town of Uglich in what many suspected was an assassination, the townspeople rose up and tolled the church bell. In a rage, Boris Godunov, who stood to benefit from the death, ordered the removal of the bell's 'tongue', had it flogged and exiled to Siberia along with the townspeople he had not already had executed. From this outrage arose stories of bells having a fondness for children, and the idea that water touched by a bell had healing powers. They also become a symbol of the underdog: the nineteenth-century progressive intellectual Alexander Herzen named his influential clandestine journal *Kolokol*: 'The Bell'.

Thanks to this potent symbolism, bells feature repeatedly at pivotal moments in Russian literature and music. In *War and Peace*, the bells of the Kremlin in Moscow ring out during Napoleon's invasion and deter the Grande Armée. In *Crime and Punishment*, the guilt-ridden Raskolnikov falls into a fever when he hears church bells ringing, and gives himself away by returning to the scene of the crime and compulsively ringing the doorbell of the old woman he murdered. Bells toll at especially dramatic points in Mussorgsky's opera *Boris Godunov*, at the climax of Tchaikovsky's 1812 Overture and in the final movement of Shostakovich's Symphony No. 11.

For the Soviets, bells were symbols of the power of the Church, and therefore of superstition and oppression. In Stalin's 'Great Turn' of the late 1920s bell-ringing was prohibited and thousands of bells were melted down. The Church of the Nativity in Moscow became a holding pen for circus lions, and the Cathedral of Christ the Saviour was replaced with a giant open-air swimming pool. Rural churches became carpentry or plumbing collectives. But religion was never quite eliminated from Soviet life and culture, and a church bell plays a central role in one of the most celebrated films ever made in the Soviet Union. *Andrei Rublev*, a biopic of the fifteenth-century icon painter, directed by Andrei Tarkovsky in 1966, begins with the flight of a primitive hot air balloon from a church tower. This slightly surreal episode is left unexplained, and the central episodes of the film follow Rublev through years of turmoil and war. But the long final section turns away from the artist to follow the casting and tolling of a huge bell. The process – finding the right clay for the mould, selecting the metal, squeezing money out of the prince, powering up the huge furnaces – is shown in painstaking detail. The climax would seem to come when the bell, ready at last, is tolled for the first time. For this Tarkovsky uses a recording made in 1963 by Mosfilm, the main Soviet film studio, of the bells of Rostov Cathedral – some of the only ones that had survived Stalin's purges. It is an astonishing sound but it is, arguably, not the most compelling one in these last scenes of the film. A few minutes before, as the molten bronze is poured, and just at the point when Boriska, the orphan who is in charge of the casting, is saying, 'O Lord help us. Let it work!', there are unearthly noises on the soundtrack that would better match Tarkovsky's 1972 science-fiction film *Solaris* than one set in late-medieval Russia.

What is going on? The bell is the bell, but it is also a symbol for much more. As the critic Petr Král writes, Tarkovsky's films are 'inseparably literal and metaphorical', and the director is, I think, enacting the mysterious transformation, both ancient and modern, that is realised in metallurgy. *Andrei Rublev* was made nearly ten years into the space age, in which the USSR had taken an early lead in 1957 with Sputnik 1, the first artificial satellite, and again in 1961 with the first human space flight by Yuri Gagarin. And for all the apparent hyper-rationalism and potential military benefit of the Soviet space programme, there was a powerful mystical and mythic drive at its heart. Sergei Korolev, the chief designer of Sputnik and Gagarin's spacecraft, the Vostok 1, had been profoundly influenced by the Cosmists, a movement of the previous generation that claimed that humanity could achieve immortality. Tarkovsky's balloon and bell, like Korolev's spacecraft, are vessels and voices of the Russian sky.

The relationship between bells and music in the West has often been an uneasy one. Even bells that have been carefully tuned retain a significant degree of inharmonicity, which means that the frequencies of overtones they produce are not the whole multiples of the fundamental frequency that our ears find harmonious. This means that, unlike a musical instrument, a bell is always 'out of tune' with itself. (One reason for this is that, unlike air columns and strings that vibrate predominantly in one dimension, bells vibrate flexurally in three dimensions.) In *The Musical Human* Michael Spitzer goes so far as to say that the overtones generated by a bell are so complex that they are practically useless for music. And yet the resonant quality of bells and what the composer David Bruce identifies as their 'immutability' has been welcomed by some musicians and deployed in remarkable ways.

Arvo Pärt incorporates a bell into his *Cantus in Memory of Benjamin Britten* (1977), a work that is simple in structure but packs a powerful emotional punch. It starts with a passing bell, tolling out of silence, which is then joined by the first violins of a string orchestra. The second violins, violas, cellos and double basses enter in turn, playing down larger and larger fragments of an eight-note scale. The effect has some resemblance to a descending sequence of English bells, but there are two differences. First, the scale is a natural minor, or Aeolian mode, rather than a major scale. Second, each voice plays the scale more slowly by the ratio of two to one so that the double basses play at a sixteenth of the speed of the first violins. Using this form – a prolation canon, which dates to the Renaissance – Pärt creates music that is at once old and new. The voices of the orchestra repeat the descending scale in a series of cascades that grow in volume until settling on a low A with all six voices playing fortissimo for thirty beats. As they end, the resonance of the bell can be heard slowly falling away again into silence.

Advances in sound engineering have opened new possibilities for the bell. For his 1980 composition *Mortuos Plango*, Jonathan Harvey sampled the sound of the great tenor bell at Winchester Cathedral, which he found to have thirty-three partials, or overtones, and manipulated and combined them with 'purely digital creations'. He then interleaved and merged these with recordings of the voice of his son, who had been a chorister at the cathedral, singing words that had been set around the mouth of the bell itself when it was cast: *Horas Avolantes Numero, Mortuos Plango, Vivos Ad Preces Voco* – 'I count the feeling hours, I lament the dead, I call the living to prayer.' The piece is designed to be replayed on eight speakers around a large auditorium. In this way, Harvey wrote, 'the ideal listener is "inside" the bell, with its partials distributed in space [around them]; the boy's voice flies around, derived from, yet becoming the bell sound'.

Mortuos Plango is a world away from Pärt's *Cantus*. It can be a little daunting when heard for the first time, and the flattening effect of a domestic sound system doesn't help. But the piece repays more than one listen. Unmoored, and floating free of normal expectation, one may begin to engage more closely with the nature of the sounds – highlighted and sea-changed by electronic manipulation such that, for instance, there is after-resonance without strike, and a transformation of the bell into the dearest sound that any of us will know: the human voice. Is this a dialogue between the living and the dead, between memory and possibility? 'The territory that the new computer technology opens up is unprecedentedly vast,' Harvey wrote;

'one is humbly aware that it will only be conquered by penetration of the human spirit, however beguiling the exhibits of technical wizardry; and that penetration will neither be rapid or easy.'

Pärt's *Cantus* lasts just five or six minutes, and Harvey's *Mortuos Plango* about nine. Others have created works for bells or bell-like instruments that are intended to last a thousand years or more. Jem Finer's *Longplayer*, for instance, began playing at midnight on 31 December 1999, and is intended to continue playing until the last moment of 2999, at which point it will begin again. In this piece, singing bowls, with their especially pure and enduring resonance, have been pre-recorded playing six short pieces of music – although on certain occasions musicians perform them live. One section from each of the six pieces is played simultaneously at all times. An algorithm chooses and combines the sections in such a way that no combination is repeated until exactly 1,000 years have passed.

Finer, who in a previous life played with The Pogues (I may never get their version of 'South Australia' out of my head or my feet), says that *Longplayer* operates rather like a system of planets 'which are aligned only once every thousand years, and whose orbits meanwhile move in and out of phase with each other in constantly shifting configurations'. He says that the preoccupations that led to its conception concerned time more than they did music. 'At extremes of scale, time has always appeared to me as baffling, both in the transience of its passing on quantum mechanical levels and in the unfathomable expanses of geological and cosmological time, in which a human lifetime is reduced to no more than a blip.' For the listener the effect is of gentle, unending cycles of overlapping resonances. At one moment it is as if one is surrounded by slowly chiming celestial clocks; at another, a meditative state in which sound is suspended somewhere between music and time itself.

The bells imagined by Brian Eno for a project called *The Clock of the Long Now* will, if created, continue to toll far beyond even *Longplayer's* horizon. The idea is that a mechanical clock housed in a mountain in Texas will keep time accurately for at least ten thousand years. According to Stewart Brand of the Long Now Foundation, who co-initiated the project, contemporary civilisation has 'a pathologically short attention span' and the clock, designed by the inventor Danny Hillis, is an example of the kind of mechanism, icon or myth that could help create what Eno calls 'a bigger here and longer now'. Critics do not tire of pointing to the irony that funding comes from Jeff Bezos, who pioneered instant online gratification with the one-click 'buy now' button on Amazon.

In 2003 Eno released an album featuring studies for the bells of *The Clock of the Long Now*. Having looked at the English practice of ringing the changes, he realised that just ten differently pitched bells would be enough to play a different sequence almost every day over the entire period (10!, or ten factorial, is 3,628,800 – very close to the number of days in 10,000 years). 'A listener on any given day,' Eno explains, 'should, if the algorithm that generates the series is known, be able to calculate the number of days since the series started playing.' The album reproduces the sequences that would play in January 7003, halfway through the first 10,000 years of the clock. In other tracks, Eno explores a hyperspace of possible bells including a 'reverse harmonic' in which the reverberation comes before the strike, and a computer-generated version of the sound of the Tsar Bell, an eighteenth-century Russian behemoth which would have been the biggest to toll had it not broken before it could be played. It's all quite ingenious, but it is also remote, abstract and, to be frank, a bit dull. And this is why, on a bright, cold day in February at just around the time that a twenty-four-hour clock would be striking thirteen, I found myself on the waterfront at Trinity Buoy Wharf in East London.

I had arrived there in a roundabout way. After my father died in 2018 I tried, and largely failed, to find time for a series of long walks which, I hoped, would help me get my own life and my grief in perspective. A volcanologist in my community choir put me in touch with an eminent geologist who helped me identify places in Britain where rocks visible to the naked eye are more or less coterminous with six great mass-extinction events in the history of life on Earth. My idea was to build a walk around each of these places in what, I joked to another friend, would be an extended act of 'psychogeology', placing the match-flare of a single life in its larger context and taking to a slightly ridiculous extreme the fashion for psychogeography.

My first great trek turned out to be a stroll on the beach at Budleigh Salterton, a picture-book Devon village of golf- and croquet-playing retirees, sandcastles and melting ice cream. There, on an unusually low tide, it is possible to see radioactive vanadium nodules from around the time of the Permian-Triassic extinction, the greatest episode of mass death in the history of life. Other walks followed, and more may still be to come. But on that February day at Trinity Buoy Wharf I was scouting for the route of a possible deep-time walk across London that would connect places associated with the mass extinction that may be unfolding in our time, the so-called Anthropocene.

To apprehend the enormity of a mass extinction you need to think beyond the time frame of a generation or two, and at Trinity Buoy Wharf you can be in the presence of Jem Finer's *Longplayer*. Its singing bowls are housed in the building where the pioneer of electricity Michael Faraday once experimented and, mounting its spiral steps, you can sit in the top of the small lighthouse to hear the bowls' thousand-year song. And here too you can visit a remarkable object mounted on the side of the wharf above the river: a bell that is tolled by the tide.

It's a funny-looking thing: a large brass bell that has some-how doubled itself, as if reflected upwards into the sky as well as pointing down to the water. Each of the two halves has two waists pinched rather like a classic Coca-Cola bottle. The odd shape is a corollary of the bell's unique sound: the founders used finite element analysis to model flexural vibrations in the prototypes, and lathes to reduce irregularities in the thickness of the walls of the finished bells to about ten-millionths of a metre. As a result, the bell has, according to its makers, some of the purest harmonics and longest resonance times of any yet made. It is rung by a long brass rod extending far down and ending in a double cone that is pushed to and fro by the rising and falling water. On the cone are raised letters in bronze reading: 'What is the song in the wave if not that all living is meeting? Nothing given or held for good.'

The bell at Trinity Buoy Wharf is one of about a dozen around the coast of Britain conceived by the sculptor Marcus Vergette and installed as part of a project called *Time and Tide*, which aims 'to celebrate and reinforce connections in local communities, between different parts of the country, between the land and sea, [and] between ourselves and our environment'. It is an idea drawing on a long history of associations between bells and the sea. After the town of Dunwich in Suffolk on the east coast of England was inundated by storm surges and disappeared beneath the sea in the fourteenth century, a legend grew that its church bells could still be heard from beneath the waves on certain tides. (Very occasionally, when the water is exceptionally clear, the church tower can still be seen looming through the water, covered in pink sponges and crawling with crabs and lobsters.) Elsewhere, from the mid nineteenth century onwards, bells were placed on buoys at sea to warn passing ships of danger, and engineers experimented with underwater bells for the same purpose at the start of the twentieth. In *The Dry Salvages*, T. S. Eliot suggests that, rung by the unhurried swell, sea bells sound a time older than the time of chronometers. The *Time and Tide* bells carry this legacy from the past, but they also sound for the future, and the great seas to come.

Resonance (2)

In 2014 the acoustic engineer Trevor Cox discovered the most resonant space in the world. It was a giant reservoir, now empty, that had been built deep inside a hill at Inchindown in Scotland in the late 1930s to store fuel oil for the Royal Navy out of reach of German bombers. The concrete walls, clogged and smoothed with oil residue, made an almost perfect surface for sound waves to bounce off, and when Cox fired a starting pistol inside the space – at 240 metres or 787 feet, more than twice the length of a football field – the reverberation continued for 112 seconds: nearly two minutes.

It was clearly an amazing experience, and Cox describes himself as whooping and jumping about like an overexcited small child before getting down to the serious business of precise measurement. But reverb like this would also make Inchindown just about the worst place in the world to perform music. Modern venues such as the Symphony Hall in Boston, which is widely judged to be one of the best, have a resonance time of around two seconds. This is enough to add depth and richness to the sound but not so much that the notes, phrases and timbres blur.

'Sounds,' the philosopher Jonathan Rée observes, 'can barely survive the moment of their creation.' But resonance enlarges and transforms them, and can seem to take the maker or listener into a dimension beyond normal experience. Our fascination with resonance may be ancient: some evidence suggests that the artists who created the oldest known cave paintings in Europe positioned them at points where echoes from the rock were strongest. At other times and places people may have deliberately sought to eliminate it. On a visit to Maes Howe, a Neolithic chambered tomb in Orkney, the poet Kathleen Jamie was struck by 'a thick soundlessness', as in a recording studio, or a strongroom. 'A moment ago, you were in the middle of a field, with the wind and curlews calling. That world has been taken away, and the world you have entered is not like a cave, but a place of artifice, of skill.'

For a thousand years after it was built in 537 CE, the cathedral of Hagia Sophia in Constantinople was the largest interior space in the world and its most resonant. Lined with marble and gold mosaic that reflected the sun and sea-light pouring through its many windows, it was luminous, and for over nine hundred years it was filled with liturgical chant. After the Muslim conquest in 1453, however, the cathedral was turned into a mosque and music was forbidden. The ban has continued ever since, but between 2010 and 2016 a team including the art historian Bissera Pentcheva, the acoustician Jonathan Abel and the choir Cappella Romana found a way to recreate the sounds that had once filled its extraordinary space. Pentcheva was allowed to pop four balloons inside the building and measure the patterns of reverberation. Abel then developed a digital filter that could imprint this sonic signature on to other sounds, including performances of Byzantine chant by the choir. You can hear the results on the podcast *The World According to Sound*. First, Cappella Romana sings in what sounds like a neutral space without echo. The music – a varying line over a drone that is then doubled at a lower octave – is simple and quietly beautiful. But then, as the reverb of Hagia Sophia is added, a new world opens up on every side.

At around ten to thirteen seconds, the resonance of Hagia Sophia is an order of magnitude (that is, ten times) shorter than Inchindown, and an order of magnitude longer than modern performance spaces such as the Boston Symphony Hall. Like the great cathedrals of western Europe, this kind of acoustic favours slow-moving musical lines. Hagia Sophia means 'holy wisdom', but the extended resonance blurs the words of the liturgy until they are almost incomprehensible, just as the play of light in its interior seems to dissolve architectural form. The building can be seen as an instrument for the human voice, but one that puts it in touch with what believers experience as the numinous. The impression, suggests Pentcheva, may be that divine knowledge can be 'grasped only partially'.

One does not have to be on board with a particular set of beliefs to recognise the value of large, resonant spaces for contemplation, ritual and music. And one could make a case that 'big resonance' has been having a moment in recent years. Among the pioneers are The Deep Listening Band, who at the end of the 1980s started recording in spaces such as giant cisterns with a reverb of up to forty-five seconds. Another example among many are the improvisations for solo saxophone by Branford Marsalis recorded in Grace Cathedral in San Francisco in 2014. And big resonance particularly suits a recent-wave choral music that is often characterised by unorthodox or crunchy harmonies. 'This transformation of dissonance into a place of comfort and refuge in contemporary choral in resonant spaces ... is an astounding inversion of the way music is supposed to work,' observes the music presenter Tom Service.

Few if any have taken experiments with resonance further than the vocal group Roomful of Teeth. Performing in 'The Tank', a steel water tower in the Colorado high desert that is twenty metres high and twelve metres wide and has a reverb of up to forty seconds, the singers transform Michael Praetorius's 1609 setting of 'Es ist ein Ros entsprungen' into 'How a Rose', a work that seems at once ancient and entirely new. Regarding resonance and its ability to both contain and expand a sound, one may recall what Fernando Pessoa wrote of Walt Whitman: 'You sang everything, and in you everything sang.'

Frontiers

Careful attention to sound can change the world. Here's an example. In the late 1500s, the composer and music theorist Vincenzo Galilei showed that the pitch of a vibrating string on an instrument such as a lute is proportional to the square root of the tension applied to it rather than, as had been assumed, the first power of that tension. This was the first demonstration of a nonlinear law in physics. No less importantly, Vincenzo's rigorous and painstaking approach and his refusal to take old ideas on trust inspired his son Galileo Galilei to develop systematic approaches to questions about natural phenomena that have helped advance scientific understanding ever since. Even today, paying attention to sound may help research at the frontiers of physics. 'If we treat music as a test ground for wave phenomena and other physics,' writes the young quantum physicist Katie McCormick, 'then the gut sense of beauty one acquires through "messing about" with music might be a useful guide to new ideas.'

There is also much to discover about how sound shapes life. 'The role that vibrations play in the physics of living systems remains largely unknown,' says the experimental physicist Shamit Shrivastava. Neurons, he points out, vibrate when they pass electrochemical signals from one end to the other. The vibration travels along the surface of the neuron, almost like a drum. 'Understanding this process could be key to innovations ranging from non-invasive treatment of neurological disorders . . . to the development of more energy-efficient computers,' he says.

The study of acoustic and seismic waves is also enriching researchers' understanding of large-scale processes in the living world. New devices that enable cheap, reliable and widespread acoustic monitoring enable biologists and ecologists to detect and monitor subtle changes, and understand behaviours and interactions that were previously hidden. One example is AudioMoth, a small field recording unit which costs only a few tens of dollars. It can be deployed and networked in large numbers to monitor continuously over extensive and hard-to-reach areas, without the need for human intervention. In Belize, AudioMoth has been used to monitor poaching of protected animals in national parks, and in Britain it has tracked the movements and behaviour of rare, elusive and poorly understood bats. HydroMoth, a version for underwater use, promises the same capability at sea, with much greater potential reach than conventional hydrophones at a fraction of the cost.

One of the most encouraging examples to date of the deployment of new acoustic technology, however, is the use of hydrophones to protect whales in the Gulf of St Lawrence on the west coast of North America. This is a busy zone for shipping, and from 2017 to 2019 dozens of right whales, who had followed their plankton food source as it moved there from warmer waters, were being killed every year as a result of ship strikes or entanglement in fishing nets. Working with government and industry, the biologist Kimberley Davies installed a network of underwater autonomous gliders equipped with hydrophones that could detect the whales' locations and direction of movement. The network transmits data in real time to ships, which can then slow down and avoid the whales. In 2020 and 2021 there were no recorded right whale deaths in the entire area.

There are new ways, too, to study sound underground. The writer Lucy Neal describes how, walking through the Białowieza Forest in Poland, she can imagine the 'chthonic hullabaloo' beneath her feet as the ants, beetles and other soil-dwellers pursue their lives. But now it is increasingly feasible not merely to imagine but to actually hear all that thrumming and scraping. In the new field of soil ecoacoustics, biologists have found that simple metal nails pushed into the earth can become upside-down antennae when connected to sensors. With these they can listen to the movements of worms, grubs, springtails, mites and many other life forms as they hunt, eat and slither past, or drum, tap and sing to get one another's attention. Even plant roots make noises as they push through soil, and by tracking these noises the soil acousticians hope to better understand hitherto unanswerable questions such as whether roots grow at day or at night, or only after rain. Monitoring the abundance and diversity of sounds made by life in the soil, or its absence, can also help determine soil health – an increasingly vital matter in a world where so much land is degraded or at risk of degradation.

Research with and into seismic waves opens other frontiers. Like sound waves, seismic waves are vibrations through matter, but unlike sound waves they take more than one form. One of these forms, called P-waves, will travel through soil, rock and other solids as well as through liquids, compressing and releasing them as they go in the same way as sound travels through air. But another type, called S-waves, move from side to side or up and down. S-waves travel more slowly than P-waves and cannot pass through liquids and gasses. A relatively simple set of equations describe the behaviour of these different waves at all scales, explains the geophysicist Tarje

Nissen-Meyer, and as a result it is usually possible to determine the nature and distribution of the different materials through which seismic waves pass and so to create a detailed and variegated three-dimensional picture of that interior, whether it is a few metres below the surface or the whole interior of a planet or moon. It means that one can determine the origin of naturally occurring seismic waves with precision. And it even means, says Nissen-Meyer, that we can know with some certainty what it would sound like to skate over the surface of Jupiter's icy moon Europa as well as model the moon's interior. More immediately, seismologists can study the movement of melting ice in Antarctica and monitor the movements of large numbers of animals in the African savannah in real time, tracking their positions to within a few metres. Researchers can also explore more deeply than ever before the extent to which the animals themselves may use seismic waves to map and communicate across the 'vibroscape'. Could it be, asks Nissen-Meyer, that elephants – creatures of great intelligence and subtlety – use them to locate sources of water hidden underground?

Other researchers seek to understand the nature of sonic communication between animals to a far greater level of detail and complexity than ever before. Yossi Yovel at Tel Aviv University, Natalie Uomini at the Max Planck Institute and Daniela Rus at MIT, for instance, are applying machine learning to follow exchanges between bats, New Caledonian crows and sperm whales respectively, each with a view to unpacking as never before what is really going on. How long will it be, wonders the writer Matthew De Abaitua, before we have a Dolittle machine, named for a character in children's fiction who is able to talk to the animals, or a real-life Babel Fish, the universal translation device in *The Hitchhiker's Guide to the Galaxy*? With devices such as these, he speculates, we may discover that the air, the earth and the water around us 'resound with the music of emotions both strange and familiar,' liberating us from 'the delusion that we are above it all,' and giving us a chance to say sorry to the species we have pushed close to extinction.

Machine learning is also expanding the nature and range of speech between humans and the machines themselves. In 2022, an engineer at Google named Blake Lemoine claimed that the company's Language Model for Dialogue Applications (or LaMDA), an AI-powered chatbot, was a sentient being. Few if any credible computer scientists or philosophers accepted his claim, but that may not be the point – which is, rather, that programmes such as LaMDA are becoming increasingly convincing for the general public and it will not be long until many people believe they are having real conversations with machines. Versions that are new at the time of writing such as ChatGPT and Bing may not last, but more sophisticated systems will almost certainly follow. 'The digitally re-enchanted world will flatter us by its seeming attentiveness to our solicitations, by its apparent anticipations of our desires, and perhaps even by its beguiling eloquence,' says the writer on technology L. M. Sacasas. The prospect may seem attractive, but Sacasas warns of dangers as the technology manipulates and shapes its users on behalf of its owners or governments. 'We are [already] surrounded by algorithms that are purportedly tailored to our preexisting preferences,' argues the writer and film-maker Noah Millman, 'but the process of being so surrounded is also training us to be algorithmically tractable.'

The music of the future will also be shaped profoundly by machine learning. It is becoming possible, for instance, to ask an AI to generate compositions by giving it a few descriptive words, just as an open AI system such as DALL-E already generates images in this way. How far could this go? 'One thing I have no fear of is that humanity will ever lose its need to be creative or the need to hear art that has been created by humans,' says the composer David Bruce. 'If anything, the pleasure of hearing human performance is likely to grow ever more precious in a world that is dominated by computers and technology.'

In all events, the number of conversations between humans and computers is likely to grow. Indeed, the technologists Benjamin Bratton and Blaise Agüera y Arcas suggest it may not be long before AI-powered systems that speak human-based language outnumber actual humans. For Bratton and Agüera y Arcas this does not have to be a bad outcome: perhaps the future can contain something like 'Golem XIV', an AI imagined by Stanisław Lem which refuses to work on military applications and other self-destructive measures, and instead is interested in the wonder and nature of the world. 'As planetary-scale computation and artificial intelligence are today often used for trivial, stupid and destructive things,' write Bratton and Agüera y Arcas, 'such a shift would be welcome and necessary.' In a way that is currently only imagined in works of science fiction such as *A Half-Built Garden* by Ruthanna Emrys, machine-learning systems might post on behalf of ecosystems, 'speaking' for trees and bacteria and soil and rivers, turning sensor data and historical trends into narratives and points of view in support of ecological restoration.

The universe, wrote Galileo Galilei, is written in the language of mathematics. But so far as we can yet determine, mathematics is not sufficient for an account of the human heart, let alone the world we co-create. Ceaseless scientific exploration has little meaning unless it helps those who explore to truly know a place, and therefore to truly value it, almost as if for the first time. And to this end we need among our tools not only numbers but also words, both old and new, and the webs of association and story that those words carry. In this regard, initiatives such as the neologisms proposed by the ecological philosopher Ginny Battson and the 'positive lexicography project' created by the psychologist Tim Lomas may help. Among Battson's new words is 'tissumble' for a 'noisy, colourful nature moment of memory and imagination, triggered by sounds'. Lomas, meanwhile, has compiled a compendium of already existing words from dozens of languages which do not have equivalents in English but which distil an experience, feeling or quality, and in doing so can help us to recognise and validate it. Some of the entries capture various special kinds of joy or appreciation. There is *utepils*, a Norwegian word for 'a beer that is enjoyed outside . . . particularly on the first hot day of the year'. There is Swahili *mbuki-mvuki*: 'to shed clothes to dance uninhibitedly'. Arabic has *tarab*: 'musically induced ecstasy or enchantment'. Tagalog has *gigil*: 'the irresistible urge to pinch or squeeze someone because they are loved and cherished'. Other entries in the lexicon describe admirable qualities of character. Among those notable for our times is *sisu*, a Finnish word for 'a consistent and courageous approach towards challenges that enables one to see beyond present limitations and into what might be'. But one word in particular stands out for me, and that is *dadirri*, an Indigenous Australian term for 'a deep, spiritual act of reflective and respectful listening'.

We also need song. In *Horizon*, his last but one book published in 2019, Barry Lopez recounts how biologists seeking to protect the mala, or rufous hare-wallaby of Western Australia, were confident that they could get the 'biological' part of a restoration project right. The mechanics of captive breeding and the selection of suitable habitat were, they said, well understood. But they felt their efforts would eventually fail because they had no knowledge of the spiritual nature of the mala and its place in the *Tjukurrpa*, or Dreaming, of the Warlpiri people for whom it is an ancestral being. So they asked Warlpiri elders to assist in the reintroduction by 'singing the wallaby up' – that is, by ritually calling mala back into the country. 'Singing' an animal back into existence might seem unscientific but, suggests Lopez, it is 'quaint only in the minds of those who believe they already know, or can discover precisely how the world is hinged'. If there is to be a future worth living in it will surely hold a place for re-enchantment.

Silence

The first thing that struck Dan Hikuroa about Antarctica was the silence. It was a windless day, and Hikuroa, who had come to the continent to study geology and fossils, remembers sitting down and hearing nothing but a faint rustling that stopped and started to a regular beat. The sound, he soon realised, came from the only thing in the vast landscape that was making a noise: a vein on his forehead brushing against his balaclava as it pulsed. His experience brings into focus two truths: very few places on Earth are almost totally silent, and even in those that are, humans bring noise with them.

The composer John Cage found something similar when he visited an anechoic chamber – the quietest possible human-made interior space – and heard what he described as two tones: a dull roar and a high whine. The first he attributed to the flow of the blood around his body; the second to the firing of his own neurons (although it may have been mild tinnitus, or the normal and natural vibration of hair bundles on cells in the inner ear: these make sound but it is so quiet that it is usually inaudible). The encounter helped inspire *4'33'*, a work for which Cage is famous today, although at least six others had created soundless works before him. In the piece, the player or players remain quiet and still for four minutes and thirty-three seconds in front of blank scores 'written' by the composer. The idea is that everyone attends to the silence around them, which turns out to be full of chance noises – whether of the wind and rain outside, people leaving the hall, or the chatter in our heads.

There are many kinds of silence. The writer and political thinker Paul Goodman came up with a poetic list of nine, including the dumb silence of slumber or apathy; the alive silence of alert perception ready to say 'this ... this ...'; the musical silence that accompanies absorbed activity; the silence of listening to another speak; the noisy silence of resentment and self-recrimination; and baffled silence. The sculptor Louise Bourgeois was also interested in its various forms: 'the length of the silence, the depth of the silence, the irony of the silence, the timing of the silence. The hostility of the silence. The shininess and the love of the silence.' One of her late works, created in 2006 a few years before her death at the age of ninety-eight, consists of just a few words written across a musical score: *Mon Dieu, Mon Dieu, Que le Silence Est Beau* (My God, My God, Silence is Beautiful). And an aspect of that beauty is captured by the critic and thinker John Berger when he describes silence as the most important thing about some masterpieces of European visual art. 'It's as if the painting – absolutely still, soundless – becomes a corridor,' he says, 'connecting the moment it represents with the moment at which you are looking at it, and something travels down that corridor at a speed greater than light, throwing into question our way of measuring time itself.'

The ultimate destination of both life and art is oblivion. And it is this kind of silence that Philip Larkin apprehends in 'Aubade', his poem of what the Anglo-Saxons called *uhtcare*, or dawn care, when he wakes at four to 'soundless dark'. Part of the condition of being human is to have the foresight to know that, no matter how solid the structures we build, no matter how vital our own lives are, they will end. 'Of a thousand years of joys and sorrows / Not a trace can be found,' wrote

Ai Qing, father of the artist Ai Weiwei, while visiting the ruins of an ancient city on the Silk Road. So, yes, we know that the houses go under the sea, and the dancers go under the hill, but even before death great chunks of our lives fall into the dark and backward abysm. 'So few things are recorded,' writes the novelist Javier Marías; 'fleeting thoughts and actions, plans and desires, secret doubts, daydreams, acts of cruelty and insults, words spoken and heard and later denied or misunderstood or distorted, promises made and then overlooked . . . how little remains of anything, and how much of that little is never talked about . . .'.

No wonder, then, that humans struggle to keep lost voices alive, and sometimes imagine them when in reality there is only silence. Haunting is, as the anthropologists Kassandra Spooner-Lockyer and Katie Kilroy-Marac write, a universal condition. 'Some times the nite is the shape of a ear only it aint a ear we know the shape of,' says the young narrator of *Riddley Walker*, Russell Hoban's novel of a distant post-apocalyptic future. 'Lissening back for all the souns whatre gone from us. The hummering of the dead towns and the voyces befor the towns ben there.' And in our times that haunting is often facilitated rather than diminished by technology. The father of a young woman killed in the attack on the Bataclan nightclub in Paris in 2015, in which ninety people were murdered and over 200 injured, was still paying his daughter's phone bill six years after her death just to hear her voicemail greeting. And the wind phone, or *kaze no denwa*, an unconnected telephone booth in Ōtsuchi in Japan, remains a channel through which the living talk to loved ones among the 15,000 who were drowned or crushed in the 2011 Tōhoku tsunami.

Sometimes the silence of death is embraced with wit and even humour. The gravestone of the composer Alfred Schnittke, for instance, shows the musical symbol denoting a rest (i.e. silence) beneath a *fermata* (indicating a pause of indefinite length), and marked *fff* – that is: make no sound for a prolonged and perhaps infinite period, and do it as loudly as possible. The rest is silence.

So long as men can breathe, or eyes can see, there will be what the philosopher Avishai Margalit calls duties of memory, and to actively silence these memories is an act of psychological violence. In what, with hindsight, was part of the preparations for large-scale physical violence with the invasion of Ukraine in a bid to erase its history, a court in Russia ordered the closure in late 2021 of the human rights group Memorial, which had been created in 1989 by the nuclear physicist and activist Andrei Sakharov and others in order to document and preserve records of atrocities of the Soviet era including the mass executions of the 1930s and the millions who suffered and died in the Gulag.

Smothering memory is seldom the whole story. In the Russian case, it was accompanied by a 'firehose of falsehood' with a view to creating conditions in which nothing is true and everything is possible, but in which fake stories will resonate with many people. And authoritarian regimes muffle present and future speech as well as memory. Among the early targets of the military in Myanmar when they seized power in 2021 were poets who opposed them. 'After the first and second poets were killed,' reported the journalist Hannah Beech, 'the third poet wrote a poem . . . After the third poet was killed, the fourth poet wrote a poem . . . After the fourth poet was killed, his body consumed by fire . . . there was no verse. At least for a moment.' But even after the most brutal acts of stifling, music and song often find a way back. In Afghanistan the Taliban have silenced musicians, but at the time of writing students and teachers from the Afghan National Institute of Music are re-establishing their school in Portugal.

Countries that are relatively open and democratic have their deliberate silencings too. To name just one among many examples in the history of empire, the British government hid evidence of large-scale torture and murder by its colonial administration in Kenya in the 1950s for decades afterwards. 'Arguably the British are not motivated by imperial nostalgia but by imperial denial,' writes the historian Charlotte L. Riley; 'this is not a remembrance, but a silence.' In the United States, the atrocity of the Tulsa massacre of 1921, which is one of the single worst incidents of racial violence in American history, was ignored and covered up for many decades. As an investigation and reconstruction a century later puts it, 'the final insult of the massacre came in the silence'.

Against such silences, there is always the frail hope that, just as the musicians of Afghanistan took up their voices and their instruments again when they were in a place of greater freedom, so the truth will out in a time of greater freedom. As the writer Eduardo Galeano put it, 'There is no silent history. No matter how much they burn it, no matter how much they break it, no matter how much they lie about it, human history refuses to keep its mouth shut.'

Raising one's voice in the face of institutional acts of silencing can be hard to do, however, and it requires perspective and courage. It is often far easier to be complicit or turn away. In 1946, Martin Niemöller, a Lutheran pastor, reflected on the atrocities committed by the Nazis and admitted, 'we preferred to keep silent'. This kind of silence, the fact of simply saying nothing, may be the easiest to fall into, and it is understandable too: in the words of the artist and activist Clint Smith it is 'the residue of fear'. But it can be one of the most troubling forms, and is not easily forgotten. 'In the end,' said Martin Luther King in 1968, 'we [in the Civil Rights Movement] will remember not the words of our enemies but the silence of our friends.'

It can be horribly easy to ignore the wrongs inflicted on humans and other beings who have little or no voice, or whose voices one fails to hear. I continue to be heedless of many of the impacts of the systems of extraction, production and pollution that make my way of life possible. Like a lot of people, I live on what the poet Wendell Berry calls 'the far side of a broken connection', and this is potentially catastrophic. The disconnect includes the soil beneath me. 'We do not see the bacterial mass residing beneath our feet, without which life as we know it would not exist,' reflects the economist Partha Dasgupta in a report on the economics of biodiversity. 'The activities they are engaged in are undertaken in silence to the human ear.' And it is also often all too easy to disregard damage inflicted on other humans because those humans seem too remote, or not really human at all. There is historical baggage here: the Europeans who colonised Australia, for example, often did not see the indigenous people *as* people and so determined the land on which they stood to be *Terra Nullius* – no man's land. A similar disregard can occur when one is deaf to the impact of present activities on future generations – when we treat their world as what the writer Jay Griffiths and, following her, the philosopher Roman Krznaric have called *Tempus Nullius* – nobody's time.

Standing in contrast to the silences of oblivion, cruelty and neglect are forms of silence that are generative and regenerative. The Christian mystic Simone Weil described a practice in which a silence 'is not an absence of sound but . . . the object of a positive sensation, more positive than that of sound'. In the Hindu and Buddhist traditions there is *Anahata*, from the Sanskrit word that translates as 'unstruck'. The idea here is a pregnant silence as a focal point for meditation, the 'soundless sound' of a ringing bowl that has not yet been touched, and by analogy an openness to possibility. Something along these lines can be found too in the application (suggested by the critic and historian B. N. Goswamy) to images by the artist Olivia Fraser of the word *chatak*. Literally meaning 'sparkle' in Hindi and Urdu, *chatak* is the 'unheard sound' a flower makes at the moment it opens. 'So much am I in tune with nature,' writes the poet Josh Malihabadi, 'that when the bud made that *chatak* sound, / I bent, drew close to it, and asked, "Was it me you were speaking to?"'

Chatak, like *Anahata*, courts paradox, although I am told there are flowers such as evening primroses that do actually make an audible pop as they unfurl their petals at nightfall. But whatever is the case, one may take a cue from Theodore of Sykeon, a seventh-century ascetic and saint, who said that a silent person is a throne of perceptiveness. Be quiet and listen. 'Sound does not exist by itself,' says the musician Daniel Barenboim; rather 'it has a permanent constant and unavoidable relation with silence. And therefore the music does not start from the first note [but] comes out of the silence that precedes it.'

Writers and poets have long stressed the importance of keeping quiet and attending to one's surroundings. 'I wish to hear the silence of the night, for the silence is something positive and to be heard,' notes Henry David Thoreau in his journal. And in *Keeping Quiet (A Callarse)*, Pablo Neruda writes that if we were not so single-minded about keeping our lives moving, and for once did nothing, then perhaps a huge silence might interrupt the sadness of never understanding ourselves and of threatening ourselves with death. Perhaps, he continues, when we are silent the earth can teach us, just as it does when everything seems dead and later proves to be alive. And this kind of silence may have the potential to stimulate ecological and political thought and action. The explorer Erling Kagge, who completed the first unsupported solo expedition on foot to the South Pole, says that the deafening silence helped him become 'more and more attentive to the world of which [one is] part'. For Dan Hikuroa, the astonishing absence of sound in Antarctica helped inspire a call he made with colleagues for the Māori principle of *kaitiakitanga*, or guardianship and stewardship, to hold more sway on the continent.

Silence need not be total to be regenerative. For the acoustic ecologist Gordon Hempton, the absence of machine noise is sufficient. In the Hoh Rainforest of the Olympic Peninsula in Washington State, where there is little besides the sound of the water, the wind and the birds, 'the whole topography of the surrounding landscape is revealed . . . in the many layers of the echo'. Silence, Hempton says, 'is not the absence of something, but the presence of everything'. Daniel Sherrell, a writer and activist, describes how on the People's Climate March in New York City in 2014 over 300,000 demonstrators held a moment of silence for lives already lost to 'the problem'. 'Here it is,' he writes, 'a colossal hush sweeping up the crowd like something is sucking all the air out of the streets . . . You can actually hear it approaching from blocks away, though of course there is nothing to hear. When the silence reaches them, people catch their breath and stand still. The quiet is heavy, almost corporeal – not the absence of sound, but the presence of something that cannot be spoken.' For Sherrell, that presence is not just one of loss, but also one of possibility, including for the child he may one day father and for whom he is writing.

Back in 1983 the theologian and social critic Ivan Illich described silence as a 'commons' under threat. By this he meant a space for shared use that is not merely a resource but also an arena in which the equal and proper voice of each man and woman can be heard. The threat, as Illich saw it, was encroachment and destruction of this arena by modern means of communication. For some critics today, such as L. M. Sacasas, this is precisely what has happened in a world in which 'every sensory experience is a commoditised resource, every lacuna a content opportunity, and every moment of silence a reason for someone else to speak louder'. As a consequence, independence of mind becomes harder and harder. 'Just as clean air makes it possible to breathe,' writes the philosopher Matthew Crawford, 'silence makes it possible to think.'

'Take the cotton / of the mind's doom-ridden chatter / out

of your ears,' writes Rumi; 'Hear the booming voice of the heavens, / the roar of fate.' But sometimes we need moments of deep quiet – not the wind, the earthquake or the fire, but the still small voice. These may occur during snowfall, as the spaces between the arms of each flake absorb most of the sound passing through the air, or even on a sunny day at the beach. Exploring the barrier island of Scolt Head off the Norfolk coast one time, I tumbled into a hollow in the sand dunes where the wind and the sound of the sea were suddenly baffled. The place was a sun trap – a walled garden in which the sea was only a distant echo. But then, as I climbed up its far side, the breeze and sound of the waves hit me full-face again, and I tumbled down a small sandy cliff to crunch on wet shells, falling out of silence back into a world alive with good noises.

Some Good Sounds

First glug of wine as you pour it from a bottle.

Sigh of a small child as they finally fall asleep, and the steady breathing that follows.

Trickle of the stream at Hodder's Combe in the Quantock Hills.

Water dripping from the branches of dwarf oaks on to the moss-covered rocks at Black-a-tor copse.

Scratch, hiss and pop on a vinyl record before the first song starts.

The song of a wren. How can it be that loud?

Thin ice cracking as my kayak pushes through.

The swoosh of 'bog bounce' when you jump on healthy wet peatland.

Louis Armstrong's trumpet intro to *West End Blues* recorded on 28 June 1928.

The crackling of a campfire made with the branches of a fallen tree on a high hill in Shropshire on an evening of a full moon.

Blatter, blunk, dibble, drisk, gagy, gally, henting, kelching, mungey, rawkey, shuckish and slappy – some of the 100 words concerning rain in Britain collected by Melissa Harrison.

The modulation from A flat to B flat major at bar 23 (*un*

peu animé) of Bruyères in the second book of Preludes by Claude Debussy.

Absolute quiet on a boulder field in the Karakoram.

A recording of Hoover the talking seal saying, 'Hello there, how are you,' 'Come over here,' and 'Get out of here' in a thick Boston accent.

Street drains gurgling busily as water trickles away under a bright sky after heavy rain.

Monito, potoroo, numbat and quoll: the names of marsupials.

Shakespeare's 65th sonnet, which begins 'Since brass, nor stone, nor earth, nor boundless sea . . .'.

Mimram, Ebble, Nadder, Wylye: the names of English chalk streams.

The wash, slap and clack of pebbles in a steep shingle beach as a wave withdraws.

Vibration of a yacht hull as it accelerates through the water under the force of the wind.

Slurp of hot, newly pressed and pasteurised apple juice as, surrounded by friends, you pour it into bottles for the winter ahead.

On a Hampshire hillside church bells echo.

Thanks

Thanks to my agent James Macdonald Lockhart, to my editor Laura Barber and the team at Granta, including Isabella Depiazzi, Christine Lo and Pru Rowlandson. Thanks also to Joseph Calamia and his colleagues at the University of Chicago Press. Thanks to Linden Lawson for copy-editing, Kate Shearman for proofreading, and David Atkinson for the index.

Thanks to: Nancy Campbell, who without knowing it spurred me to actually get down to work; Gavin Francis, who talked to me about auscultation; Adrian Freedman, for describing his shakuhachi practice; Milton Garcés, who shared the sounds of volcanoes; Unto Laine, for the sounds of northern lights; Andri Snær Magnason, for reflections on an unquiet Earth; Myele Manzanza, for insight into rhythm; Tarje Nissen-Meyer, who helped me to understand a little about seismic phenomena; Philip Read and my brothers in Manchoir, for many years of singing together; Guillermo Rosenthuler, for helping me to find my head voice; Shamit Shrivastavam, for a bracing walk and talk about sound and the physics of life; Marcus Vergette, for reflecting on his work and the past, future and possible space of bells; and Ben Willmore, who first set me straight on some of the science of hearing.

Thanks to many others who, in ways big and small, informed,

inspired or encouraged me. They include: Matthew Adams, Eduardo Aladro Vico, Almudena Alonso Herrero, James Attlee, Kimberly Arcand, Anthony Barnett, James Bradley, Ben Brubaker, Susan Canney, Sonia Contera, Jo Cartmell, Tim Dee, Nick Drake, Mahan Esfahani, Olivia Fraser, Jez riley French, Charles Foster, Hiruni Samadi Galpayage, Chris Goodall, Loren Griffith, Haleh Liza Gafori, Jeremy Gilbert, Peter Gingold, David George Haskell, Tim Harford, Judith Herrin, Roland Hodson, Helen Jukes, Bill Janus, John Kitching, Roman Krznaric, Michel Lara, Antonia Layard, Laura Lorson, Robert Macfarlane, Niki Mardas, James Marriott, Fran Monks, Pedro Moura Costa, Vicent J. Martínez, George Monbiot, Gregory Norminton, Sam Parkin, David Pyle, Ben Rawlence, Kate Raworth, Henry Rothwell, Jonathan Rowson, Matt Russo, John Sims, Dominic Stichbury, Ian Tatum, Maya Tudor, Mark Vernon, Hugh Warwick, and Rebecca Wragg Sykes.

Thanks to Cristina and Lara. Thanks to my American and Spanish families.

In memory of my mother and father.

Um tom para todos nós.

References and Further Reading

Epigraph

Rumi, trans. Haleh Liza Gafori, 2022, *Gold*, New York Review of Books

Introduction

al-Quḍāt, ʿAyn, 1120, *The Essence of Reality: A Defence of Philosophical Sufism*, ed. and trans. Mohammed Rustom, 2022, NYU Press

Bakker, Karen, 2022, *The Sounds of Life: How Digital Technology Is Bringing Us Closer to the Worlds of Animals and Plants*, Princeton University Press

Barton, Adriana, 2022, *Wired for Music: A Search for Health and Joy Through the Science of Sound*, Greystone Books

Haskell, David George, 2022, *Sounds Wild and Broken: Sonic Marvels, Evolution's Creativity, and the Crisis of Sensory Extinction*, Viking

Hendy, David, 2013, *Noise: A Human History of Sound and Listening*, Profile

Hill, Don, 'Listening to Stones: Learning in Leroy Little Bear's Laboratory: Dialogue in the World Outside', *Alberta Views*, 1 September 2008, https://albertaviews.ca/listening-to-stones/

Kemp, Luke, et al., 'Climate Endgame: Exploring Catastrophic Climate Change Scenarios', *PNAS*, 1 August 2022

Ritchie, Hannah, and Roser, Max, 2021, 'Extinctions', https://ourworldindata.org/extinctions

Rogers, Jude, 2022, *The Sound of Being Human: How Music Shapes Our Lives*, White Rabbit

Waddington, Elizabeth (undated), 'Soundscape and Acoustic Ecology: The Music of a Changing World', earth.fm https://earth.fm/details/soundscape-acoustic-ecology/

World Wildlife Fund, 2022, 'Wildlife Populations Plummet by 69%: Living Planet Report 2022', https://www.worldwildlife.org/pages/living-planet-report-2022

Yong, Ed, 2022, *An Immense World: How Animal Senses Reveal the Hidden Realms Around Us*, Bodley Head

First Sounds

Britt, Robert Roy, 2005, 'First Sound Waves Left Imprint on the Universe', https://www.space.com/661-sound-waves-left-imprint-universe.html

'La Sinfonia Cósmica – The Cosmic Symphony English subtitles', *Conec Magazine*, https://www.youtube.com/watch?v=DVgN3lGgWUc

Penrose, Roger, 'Why Did the Universe Begin?', *Aeon*, 5 November 2020, https://aeon.co/videos/a-cyclical-forgetful-universe-roger-penrose-details-an-astonishing-origin-hypothesis

Whitehead, Nadia, 'A Glimpse into the Universe's First Light', *The Harvard Gazette*, 24 March 2022, https://news.harvard.edu/gazette/story/2022/03/simulations-show-formation-of-universes-first-light/

Resonance (1)

Bleck-Neuhaus, Jörn, 'Mechanical Resonance: 300 Years from Discovery to the Full Understanding of its Importance', *Arxiv*, 20 November 2018, https://arxiv.org/abs/1811.08353

Brubaker, Ben, 'How the Physics of Resonance Shapes Reality', *Quanta Magazine*, 26 January

2022, https://www.quantamagazine.org/
   how-the-physics-of-resonance-shapes-reality-20220126/
Brubaker, Ben, 'Of Lifetimes and Linewidths' (blog
   post), 26 January 2022, https://benbrubaker.com/
   of-lifetimes-and-linewidths/
Wolchover, Natalie, 'A Primordial Nucleus Behind the
   Elements of Life', *Quanta Magazine*, 4 December
   2012, https://www.quantamagazine.org/
   the-physics-behind-the-elements-of-life-20121204/

Sound in Space

Cunio, Kim, 2020, 'Jezero Crater', from *Celestial Incantations*
   by Sounds of Space Project, https://soundsofspaceproject.
   bandcamp.com/album/celestial-incantations
Edds, Kevin, 'Space', Twenty Thousand Hertz (podcast). Also
   Melodysheep, 'The Sounds of Space: A Sonic Adventure to
   Other Worlds', 16 June 2021, https://www.youtube.com/
   watch?v=OeYnV9zp7Dk&t=408s
Holmes, Richard, 2013, *Falling Upwards: How We Took to The Air*,
   William Collins
NASA, 2021, 'Audio from Perseverance', https://www.jpl.nasa.
   gov/news/nasas-perseverance-captures-video-audio-of-fourth-
   ingenuity-flight
NASA, 2018, 'Sounds of the Sun', https://www.nasa.gov/feature/
   goddard/2018/sounds-of-the-sun
NASA (undated), 'In Depth: Titan', 'Solar System Exploration',
   https://solarsystem.nasa.gov/moons/saturn-moons/titan/
   in-depth/
NASA, 'New NASA Black Hole Sonifications with a Remix', 4 May
   2022, https://www.nasa.gov/mission_pages/chandra/news/
   new-nasa-black-hole-sonifications-with-a-remix.html
O'Callaghan, Jonathan, 'Moonquakes and Marsquakes: How We
   Peer Inside Other Worlds', *Horizon*, 10 August 2020, https://
   horizon-magazine.eu/article/moonquakes-and-marsquakes-
   how-we-peer-inside-other-worlds.html
Scharping, Nathaniel, 'What Would the Sun Sound Like if We

Could Hear It on Earth?', *Discover Magazine*, 4 February
2020, https://www.discovermagazine.com/the-sciences/
what-would-the-sun-sound-like-if-we-could-hear-it-on-earth

Shatner, William, 'My Trip to Space Filled Me with
"Overwhelming Sadness"', *Variety*, 6 October 2022, https://
variety.com/2022/tv/news/william-shatner-space-boldly-go-
excerpt-1235395113/. Shatner wrote: 'In [the] insignificance we
share, we have one gift that other species perhaps do not: we are
aware – not only of our insignificance, but the grandeur around
us that *makes* us insignificant. That allows us perhaps a chance
to rededicate ourselves to our planet, to each other, to life and
love all around us. If we seize that chance.'

Stähler, Simon C., et al., 'Seismic Wave Propagation in Icy Ocean
Worlds', 9 May 2017, https://arxiv.org/abs/1705.03500

University of Southampton, 'The Sounds of Mars and Venus Are
Revealed for the First Time', Phys.org, 2 April 2012, https://phys.
org/news/2012-04-mars-venus-revealed.html

'Want to Know What's Inside a Star? Listen Closely',
The Economist, 14 September 2022, https://www.
economist.com/science-and-technology/2022/09/14/
want-to-know-whats-inside-a-star-listen-closely

Weltevrede, Patrick, 'Joy Division: 40 Years on from Unknown
Pleasures, Astronomers Have Revisited the Pulsar from the
Iconic Album Cover', *The Conversation*, 11 July 2019, https://
theconversation.com/joy-division-40-years-on-from-unknown-
pleasures-astronomers-have-revisited-the-pulsar-from-the-
iconic-album-cover-119861

Wu, Katherine, 2018, 'If You Were Able to Talk on Another Planet,
How Would You Sound?', SITN, Harvard University, https://
sitn.hms.harvard.edu/flash/2018/talk-another-planet-sound/

Music of the Spheres (1)

Brotton, Jerry, 'Harmony of the Spheres', BBC Radio 3, 28 August
2020, https://www.bbc.co.uk/sounds/play/m0003sgc

Clark, Stuart, 'The Music of the Spheres', The Essay, BBC Radio 3,
October 2018, https://www.bbc.co.uk/programmes/m0000kgq

Digges, Thomas, 1576, translation of Copernicus's *De revolutionibus orbium coelestium*, imagines the harmony of the universe as an 'immovable ... palace of foelicitye garnished with perpetuall shininge glorious lights innumerable ... the very court of coelestiall angelles devoide of griefe and replenished with perfite endless ioye'.

Music of the Spheres (2)

Arcand, Kimberly, Russo, Matt, et al., 'Sounds from Around the Milky Way', Chandra X-Ray Observatory, Harvard-Smithsonian Center for Astrophysics, NASA's Universe of Learning Program, 21 September 2020, https://chandra.si.edu/blog/node/770

Basinski, William, 'On Time Out of Time', 8 March 2019, https://www.mmlxii.com/products/638576-on-time-out-of-time

Crumb, George, 1977, *Star Child: A Parable for Soprano, Antiphonal Children's Voices, Male Speaking Choir and Bell Ringers, and Large Orchestra*, C. F. Peters

Díaz-Merced, Wanda L., 2013, 'Sound for the Exploration of Space Physics Data', PhD thesis, University of Glasgow, https://theses.gla.ac.uk/5804/

Díaz-Merced, Wanda L., 2016, 'How a Blind Astronomer Found a Way to Hear the Stars', TED Talk, https://www.ted.com/talks/wanda_diaz_merced_how_a_blind_astronomer_found_a_way_to_hear_the_stars. See also 'Celebrating Scientists with Disabilities', Royal Society, https://royalsociety.org/topics-policy/diversity-in-science/scientists-with-disabilities/

Finer, Jem, 1999, *Longplayer*, https://longplayer.org/about/overview/

Keum, Tae-Yeoun, 2020, *Plato and the Mythic Tradition in Political Thought*, Harvard University Press, and 'Why Philosophy Needs Myth', *Aeon*, 8 November 2021, https://aeon.co/essays/was-plato-a-mythmaker-or-the-mythbuster-of-western-thought

'"Music" of Planets Is Created at Yale to Prove Theory', *New York Times*, 22 March 1977, https://www.nytimes.com/1977/03/22/archives/music-of-planets-is-created-at-yale-to-prove-theory-music-of.html

Overstreet, Wylie, and Gorosh, Alex, 2015, 'To Scale:

The Solar System', https://www.youtube.com/ watch?v=zR3Igc3Rhfg&ab_channel=ToScale%3A

The Solar System', https://www.youtube.com/
watch?v=zR3Igc3Rhfg&ab_channel=ToScale%3A

Richter, Max, 2020, *CP1919*, https://www.maxrichtermusic.com/
albums/journey-cp1919-aurora-orchestra/

Rodgers, John, and Ruff, Willie, 1979, 'Kepler's Harmony of the
World: A Realization for the Ear: Three and a half centuries
after their conception, Kepler's data plotting the harmonic
movement of the planets have been realized in sound with the
help of modern astronomical knowledge and a computer-sound
synthesizer', *American Scientist*, 67(3). See also 'The Harmony of
the World: A Realization for the Ear of Johannes Kepler's Data
from Harmonices Mundi 1619', https://www.willieruff.com/
harmony-of-the-world.html

Russo, Matt, with Santaguida, Andrew, and Tamayo, Dan, 2018,
'The Sound of Jupiter's Moons', 'Trappist Sounds', 'K2-138',
https://www.system-sounds.com/k2-138/

Russo, Matt, 2018, 'What Does the Universe Sound like? A Musical
Tour', TED Talk, https://www.ted.com/talks/matt_russo_
what_does_the_universe_sound_like_a_musical_tour

Saariaho, Kaija, 2005, *Asteroid 4179: Toutatis*, Saariaho.org https://
saariaho.org/works/asteroid-4179-toutatis/

Tamayo, Daniel, et al., 2017, 'Convergent Migration Renders
TRAPPIST-1 Long-lived', *The Astrophysical Journal Letters*, 840
L19, https://iopscience.iop.org/article/10.3847/2041-8213/aa70ea

Weinberger, Eliot, 2020, *Angels & Saints*, Christine Burgin/New
Directions

Wishart, Trevor, 2017, 'Supernova', *The Secret Resonance of
Things*, https://icrdistribution.bandcamp.com/album/
the-secret-resonance-of-things

The Golden Record

Dowland, John, 1603, 'Time Stands Still', The Third and Last Booke
of Songs or Ayres, no. 2, arr. Nico Muhly (2018), Rose Music
Publishing

'Golden Record 2.0', Science Friday, 7 October 2016, https://www.
sciencefriday.com/segments/golden-record-2-0/

Juchau, Mireille, 'What Should We Send into Space as a *New* Record of Humanity?', lithub.com, 22 April 2019, https://lithub.com/what-should-we-send-into-space-as-a-new-record-of-humanity/

NASA (undated), 'What are the Contents of the Golden Record?', https://voyager.jpl.nasa.gov/golden-record/whats-on-the-record/

Popova, Maria, 'We Are Singing Stardust: Carl Sagan on the Story of Humanity's Greatest Message and How the Golden Record Was Born', *The Marginalian*, 2 October 2014, https://www.themarginalian.org/2014/02/10/murmurs-of-earth-sagan-golden-record/

Radiolab, 'Space', radiolab.org, 6 April 2020, https://www.wnycstudios.org/story/91520-space

Spiegel, Laurie, 1977, 'Kepler's Harmony of the Worlds', https://pitchfork.com/features/article/9002-laurie-spiegel/

Taylor, Dallas, 2019. 'Voyager Golden Record', Twenty Thousand Hertz, https://www.20k.org/episodes/voyagergoldenrecord

'Your Record: We Asked You to Tell Us What You'd Put on a New Golden Record. Here's What You Chose', Science Friday, 2016, https://apps.sciencefriday.com/goldenrecord/

Rhythm (1) – Planet Waves

Barletta, Vincent, 2020, *Rhythm: Form and Dispossession*, University of Chicago Press

Berthold, Daniel, 'Aldo Leopold: In Search of a Poetic Science', *Human Ecology Review*, 11(3), 2004, pp. 205–14

Finzi, Gerald, 1936, *Earth and Air and Rain*, Op. 15 No.10, Roderick Williams and Iain Burnside

Geddes, Linda, 2019, *Chasing the Sun: The New Science of Sunlight and How It Shapes Our Bodies and Minds*, Pegasus Books

Hamilton, Andy, Paddison, Max, and Cheyne, Peter (eds), 2019, *The Philosophy of Rhythm: Aesthetics, Music, Poetics*, Oxford University Press

Hempton, Gordon, 'The Ocean is a Drum', 8 November 2016, https://www.soundtracker.com/products/the-ocean-is-a-drum/

Leopold, Aldo, 1949, *A Sand County Almanac*, Oxford University Press

Lidén, Signe, 2019, 'The Tidal Sense', exhibition at Ramberg, Lofoten, https://signeliden.com/?p=1994

Lidén, Signe, 'The Tidal Sense', A Reduced Listening production for BBC Radio 3, 21 March 2021, https://www.bbc.co.uk/programmes/mooosycw

Lopez, Barry, 1986, *Arctic Dreams: Imagination and Desire in a Northern Landscape*, Scribner's

Miłosz, Czesław, essay on *Exiles* by Josef Koudelka, June 2006, https://americansuburbx.com/2009/06/theory-czeslaw-milosz-on-josef.html

Nicolson, Adam, 2021, *The Sea Is Not Made of Water*, William Collins

Rawls, Christina, 'A Philosophy of Sound', *Aeon*, 13 April 2021, https://aeon.co/essays/the-universal-forces-of-sound-and-rhythm-enhance-thought-and-feeling

The Loudest Sound

Black, Riley, 2022, *The Last Days of the Dinosaurs: An Asteroid, Extinction, and the Beginning of Our World*, St Martin's Press

Brannen, Peter, 2018, *The Ends of the World: Volcanic Apocalypses, Lethal Oceans and Our Quest to Understand Earth's Past Mass Extinctions*, Oneworld

Benn, Jordan, 2022, 'How Loud Can Sound Physically Get?', https://m.youtube.com/watch?v=tONF9OSUOSw

The Northern Lights

Hambling, David, 'The Northern Lights make a mysterious noise and now we might know why', *New Scientist*, 3 April 2019, https://www.newscientist.com/article/mg24232240-400-the-northern-lights-make-a-mysterious-noise-and-now-we-might-know-why/

Laine, Unto, personal communication

Quark expeditions, 'Of Legends and Folklore: Greenland's Northern Lights', https://explore.quarkexpeditions.com/blog/of-legends-and-folklore-greenland-s-northern-lights

Volcano

Barras, Colin, 'Is an Aboriginal Tale of an Ancient Volcano the
  Oldest Story Ever Told?', *Science*, 11 February 2022
French, Jez riley, 'Audible Silence – a Personal Reflection on
  Listening to Sounds Outside of Our Attention', World Listening
  Project, 24 June 202, https://www.worldlisteningproject.org/
  audible-silence-a-personal-reflection-on-listening-to-sounds-
  outside-of-our-attention/
Garcés, Milton (undated), Arenal tremor and Kipu 0704221333,
  https://soundstudiesblog.com/milton-garces/
https://www.isla.hawaii.edu/sounds/earth-sounds/
Herzog, Werner, 2016, *Into the Inferno*, Netflix
Hutchison, A. A., et al., 'The 1717 Eruption of Volcan de Fuego,
  Guatemala: Cascading Hazards and Societal Response',
  Quaternary International, 394, 11 February 2016
Johnson, Jeffrey B., and Watson, Leighton M., 'Monitoring
  Volcanic Craters with Infrasound "Music"', *Eos*,
  17 June 2019, https://eos.org/science-updates/
  monitoring-volcanic-craters-with-infrasound-music
Julavits, Heidi, 'Chasing the Lava Flow in Iceland', *The New
  Yorker*, 23 August 2021, https://www.newyorker.com/
  magazine/2021/08/23/chasing-the-lava-flow-in-iceland
Magnason, Andri Snaer, 'Night Walk to Welcome our Newborn
  Volcano', andrimagnason.com, 23 March 2021, http://www.
  andrimagnason.com/news/2021/03/night-walk-to-welcome-our-
  newborn-volcano-a-short-travel-story/
Magnason, Andri Snaer, 'The Gods Were Right', *The Atlantic*, 2
  June 2021, https://www.theatlantic.com/ideas/archive/2021/06/
  iceland-volcano-carbon-eruption/619047/
Magnason, Andri Snaer, 'Kali Doing Her Thing', Twitter,
  7 June 2021, https://twitter.com/AndriMagnason/
  status/1402020675905261568
Marletto, Chiara, 'Our Little Life Is Rounded with
  Possibility', *Nautilus*, 9 June 2021, https://nautil.us/
  our-little-life-is-rounded-with-possibility-238220/
Pyle, David, 2013, 'Professor John Barry Dawson',

Volcanic Degassing (blog), https://blogs.egu.
eu/network/volcanicdegassing/2013/02/08/
professor-john-barry-dawson-1932-2013/
Seismic Sound Lab, Lamont Doherty Earth Observatory, http://
www.seismicsoundlab.org/
Wright, Corwin, et al., 'Tonga Eruption Triggered Waves
Propagating Globally from Surface to Edge of Space', Earth and
Space Science Open Archive, 3 March 2022, https://www.essoar.
org/doi/10.1002/essoar.10510674.1

Listening to a Rainbow

Blum, Dani, 'Can Brown Noise Turn Off Your Brain?', *New
York Times*, 23 September 2022, https://www.nytimes.com/
interactive/2022/09/23/well/mind/brown-noise.html
Cleeves, L. Ilsedore, et al., 'The Ancient Heritage of Water Ice in
the Solar System', *Science*, 345(6204), 2014
Dunn, Douglas, 1985, 'A Rediscovery of Juvenilia', in *Elegies*, Faber
& Faber
Dunn, Douglas, 2019, 'Wondrous Strange', in *The Noise of a Fly*,
Faber & Faber
Neal, Meghan, 'The Many Colors of Sound', *The Atlantic*, 16
February 2016, https://www.theatlantic.com/science/
archive/2016/02/white-noise-sound-colors/462972/
Papalambros, Nelly A., et al., 2017, 'Acoustic Enhancement of Sleep
Slow Oscillations and Concomitant Memory Improvement in
Older Adults', *Frontiers in Human Neuroscience*, 8 March 2017

Rhythm (2) – Body

Buzsáki, György, 2019, *The Brain from Inside Out In*, Oxford
University Press
Nestor, James, 2020, *Breath: The New Science of a Lost Art*, Riverhead
Books
Rilke, Rainer Maria, trans. Don Paterson, 2006, *Orpheus, A Version
of Rilke*, Faber & Faber

Hearing

Arora, Nikita, 'A Touch of Moss', *Aeon*, 8 September 2022, https://aeon.co/essays/a-history-of-botany-and-colonialism-touched-off-by-a-moss-bed

Ashby, Jack, 2022, *Platypus Matters: The Extraordinary Story of Australian Mammals*, HarperCollins

Bathhurst, Bella, 2017, *Sound: A Story of Hearing Lost and Found*, Profile

Blundon, Elizabeth G., et al., 2020, 'Electrophysiological Evidence of Preserved Hearing at the End of Life', *Nature*, 25 June 2020

Bradley, James, 'Do Fish Dream?', *Cosmos*, 9 December 2020, https://cosmosmagazine.com/nature/marine-life/do-fish-dream/

Burnside, John, 'The Inner Ear', *London Review of Books*, 13 December 2001, https://www.lrb.co.uk/the-paper/v23/n24/john-burnside/the-inner-ear

Christensen, C. B., et al., 2015, 'Better than Fish on Land? Hearing Across Metamorphosis in Salamanders', *Proceedings of the Royal Society B*, 7 March 2015

Clack, Jennifer, 'The Origin of Terrestrial Hearing', *Nature*, 519, 2015, pp. 168–9

Dallos, P., and Fakler, B., 'Prestin, a New Type of Motor Protein', *Nature Reviews Molecular Cell Biology*, 3, 1 February 2002, pp. 104–11

Glennie, Evelyn, 2015, 'Hearing Essay', at https://www.evelyn.co.uk/hearing-essay/

Godfrey-Smith, Peter, 2020, *Metazoa: Animal Minds and the Birth of Consciousness*, William Collins

Groh, Jennifer M., 2014, *Making Space: How the Brain Knows Where Things Are*, Belknap Harvard

Hawkes, Jacquetta, (1951) 2012, *A Land*, Collins Nature Library

Hudspeth, A. J., 'The Energetic Ear', *Daedalus*, 144(1), 2015

Hudspeth, A. J., 2019, 'The Beautiful, Mysterious Science of How you Hear', TED Talk @NAS, https://www.ted.com/talks/jim_hudspeth_the_beautiful_mysterious_science_of_how_you_hear

Knight, K., 'Lungfish Hear Air-borne Sound', *Journal of Experimental Biology*, 1 February 2015

Long, John, 'Now Listen', *The Conversation*, 23 January 2014, https://theconversation.com/now-listen-air-breathing-fish-gave-humans-the-ability-to-hear-21324

Monbiot, George, 2017, *Out of the Wreckage: A New Politics for an Age of Crisis*, Verso

Pomeroy, Ross, 'There's an Amazing Reason Why Races Are Started with Gunshots', *Real Clear Science*, 3 August 2016, https://www.realclearscience.com/blog/2016/08/theres_an_amazing_reason_why_races_are_started_with_guns.html

Ancient Animal Noises

Clarke, J., et al., 'Fossil Evidence of the Avian Vocal Organ from the Mesozoic', *Nature*, 538, 12 October 2016

Darwin, Charles, 1839, *The Voyage of the Beagle*, https://www.gutenberg.org/files/944/944-h/944-h.htm

Diegert, C. F., and Williamson, T. E., 'A Digital Acoustic Model of the Lambeosaurine Hadrosaur *Parasaurolophus tubicen*', *Journal of Vertebrate Paleontology*, January 1998

Geisel, Theodor Seuss, 1962, *Dr Seuss's Sleep Book*, Random House

Gibson, Graeme, 2005, *The Bedside Book of Birds*, Doubleday Canada

Gu, Jun-Jie, et al., 'Wing Stridulation in a Jurassic Katydid Produced Low-pitched Musical Calls to Attract Females', *PNAS*, 6 February 2012

Jorgewich-Cohen, Gabriel, et al., 'Common Evolutionary Origin of Acoustic Communication in Choanate Vertebrates', *Nature Communications*, 25 October 2022

Katsnelson, Alla, 2016, 'Fossilized Cricket Song Brought to Life in a Work of Art', *PNAS*, 30 August 2016

Low, Tim, 2014, *Where Song Began: Australia's Birds and How They Changed the World*, Penguin

Merwin, W. S., 'The Laughing Thrush', https://merwinconservancy.org/2017/09/the-laughing-thrush-by-ws-merwin/

Warshall, Peter, 1999, 'Two Billion Years of Animal Sounds' (lecture), Allen Ginsberg Library and Naropa University

Archives, http://archives.naropa.edu/digital/collection/
p16621coll1/id/1347

Yong, Ed, 'The Story of Songbirds Is a Story of Sugar', *The
Atlantic*, 8 July 2021, https://www.theatlantic.com/science/
archive/2021/07/origin-of-birdsong-sugar/619387/

Plant

Appel, H. M., and Cocroft, R. B., 'Plants Respond to Leaf Vibrations
Caused by Insect Herbivore Chewing', *Oecologia* 175, 2014

Baker, J. A., (1969) 2015, *The Hill of Summer*, in *The Complete Works of
J. A. Baker*, William Collins

Beresford-Kroeger, Diana, 2010, *The Global Forest: Forty Ways Trees
Can Save Us*, Viking Penguin

Bonfante, P., and Genre, A., 'Mechanisms Underlying Beneficial
Plant–Fungus Interactions in Mycorrhizal Symbiosis', *Nature
Communications*, 1(48), 27 July 2010

French, Jez riley, 2021, 'Audible Silence – a Personal Reflection on
Listening to Sounds Outside of Our Attention', World Listening
Project, 24 June 2021, https://www.worldlisteningproject.
org/audible-silence-a-personal-reflection-on-listening-to-sounds-
outside-of-our-attention/

Khait, I., et al., 'Sound Perception in Plants', *Seminars in Cell and
Developmental Biology*, 2, August 2019

Levertov, Denise, 2013, 'A Tree Telling of Orpheus', in *Collected
Poems*, New Directions

Rawlence, Ben, 2022, *The Treeline: The Last Forest and the Future of
Life on Earth*, St Martin's Publishing Group

Shivanna, K. R., 'Phytoacoustics – Plants Can Perceive Ambient Sound
and Respond', *The Journal of the Indian Botanical Society*, 102(1), 2022

Insect

Ball, Lawrence, et al., 'The Bugs Matter Citizen Science Survey:
Counting Insect "Splats" on Vehicle Number Plates Reveals a
58.5 Per Cent Reduction in Abundance of Flying Insects in the

UK Between 2004 and 2021', *Buglife*, 5 May 2022, https://www.
buglife.org.uk/news/bugs-matter-survey-finds-that-uk-flying-
insects-have-declined-by-nearly-60-in-less-than-20-years/
Goulson, Dave, 2021, *Silent Earth: Averting the Insect Apocalypse*, Vintage
Hallmann, Caspar A., et al., 'More than 75 Per Cent Decline Over
27 Years in Total Flying Insect Biomass in Protected Areas',
PLOS 1, 18 October 2017
Sánchez-Bayo, Francisco, et al., 'Worldwide Decline of the
Entomofauna: A Review of Its Drivers', *Biological Conservation*,
April 2019

Bee

Chittka, Lars, 2022, *The Mind of a Bee*, Princeton University Press
Galpayage Dona, H. S., et al., 2022, 'Do Bumble Bees Play?', *Animal
Behaviour*, 19 October 2022
Meek, James, 'Schlepping Around the Flowers', review of *The Hive:
The Story of the Honey-Bee and Us* by Bee Wilson, *London Review
of Books*, 4 November 2004, https://www.lrb.co.uk/the-paper/
v26/n21/james-meek/schlepping-around-the-flowers
White, Gilbert, (1789) 2014, *The Natural History and Antiquities of
Selborne*, with an Introduction by James Lovelock, Little Toller

Frog

'Ariana Grande Kept Awake by Frog Sex', News24.com,
22 July 2014, https://www.news24.com/you/Archive/
ariana-grande-kept-awake-by-frog-sex-20170728
Aristophanes, 405 BC, *The Frogs*
Brunner, Rebecca, et al., 'Nocturnal Visual Displays and Call
Description of the Cascade Specialist Glassfrog *Sachatamia
orejuela*', *Behaviour*, 12 November 2020
Lee, N., et al., 'Lungs Contribute to Solving the Frog's Cocktail
Party Problem by Enhancing the Spectral Contrast of
Conspecific Vocal Signals', bioRxiv, 1 July 2020, https://doi.
org/10.1101/2020.06.30.171991

Orwell, George, 1946, 'Some Thoughts on the Common Toad',
    orwellfoundation.org
Pascoal, Hermeto, 2013, 'Hermeto e os sapos', https://
    www.youtube.com/watch?v=iFGTQDD09sc&ab_
    channel=CBMTijuca1; and Gioia, Ted, 'The Most
    Musical Man in the World Turns 85', The Honest Broker
    (blog), 21 June 2021, https://tedgioia.substack.com/p/
    the-most-musical-man-in-the-world
Warshall, Peter, 1999, 'Two Billion Years of Animal Sounds' (lecture),
    Allen Ginsberg Library and Naropa University Archives
Yovanovich, Carola A. M., et al., 'The Dual Rod System of
    Amphibians Supports Colour Discrimination at the Absolute
    Visual Threshold', *Philosophical Transactions of the Royal Society
    B*, 5 April 2017

Bat

Asma, Stephen T., 2017, *The Evolution of Imagination*, University of
    Chicago Press
Damasio, Antonio, 2021, *Feeling & Knowing: Making Minds Conscious*,
    Robinson
Dunitz, Jack D., and Joyce, Gerald F., 2013, *Leslie Orgel: A
    Biographical Memoir*, National Academy of Sciences
Håkansson, Jonas et al., 'Bats Expand Their Vocal Range by
    Recruiting Different Laryngeal Structures for Echolocation and
    Social Communication', *PLOS Biology*, 29 November 2022
Yong, Ed, 2022, *An Immense World: How Animal Senses Reveal the
    Hidden Realms Around Us*, Bodley Head

Elephant

Elephant Listening Project, K. Lisa Yang Center for
    Conservation Bioacoustics, Cornell University, https://
    elephantlisteningproject.org/
Ledgard, J. M., 2020, 'Dugong', *Alexander*, https://alxr.com/
McComb, Karen, et al., 'Long-Distance Communication of

Acoustic Cues to Social Identity in African Elephants', *Animal Behaviour*, 65, February 2003

McComb, Karen, et al., 2014, 'Elephants Can Determine Ethnicity, Gender, and Age from Acoustic Cues in Human Voices', *PNAS*, 10 March 2014

Ritchie, Hannah, and Roser, Max, 'Biodiversity/Mammals/ Elephants', Our World in Data, https://ourworldindata.org/ mammals#elephants

The Thousand-mile Song of the Whale

Adams, Matthew, 2020, 'Between the Whale and the Kāuri Tree', in *Anthropocene Psychology: Being Human in a More-Than-Human World*, Routledge

French, Kristen, 'The Mystery of the Blue Whale Songs', *Nautilus*, 23 November 2022, https://nautil.us/ the-mystery-of-the-blue-whale-songs-248099/

Giggs, Rebecca, 2020, *Fathoms: The World in the Whale*, Scribe

Hsu, Jeremy, 'The Military Wants to Hide Covert Messages in Marine Mammal Sounds', *Haika Magazine*, 10 December 2020, https://hakaimagazine.com/news/the-military-wants-to-hide-covert-messages-in-marine-mammal-sounds/

Hutson, Matthew, 'How a Marine Biologist Remixed Whalesong', *The New Yorker*, 29 November 2022, https://www.newyorker.com/science/elements/ how-a-marine-biologist-remixed-whalesong

Kolbert, Elizabeth, 'The Strange and Secret Ways That Animals Perceive the World', *The New Yorker*, 6 June 2022, https://www.newyorker.com/magazine/2022/06/13/ the-strange-and-secret-ways-that-animals-perceive-the-world-ed-yong-immense-world-tom-mustill-how-to-speak-whale

Langlois, Krista, 'When Whales and Humans Talk: Arctic People Have Been Communicating with Cetaceans for Centuries – and Scientists Are Finally Taking Note', *Haika Magazine*, 3 April 2018, https://hakaimagazine.com/features/ when-whales-and-humans-talk/

New Songs of the Humpback Whale, 2015, recordings by Salvatore

Cerchio, Oliver Adam, Glenn Edney and David Rothenberg, https://importantrecords.com/products/imprec433

Payne, Roger, 1970, *Songs of the Humpback Whale*, Capitol Records

Payne, Roger S., and McVay, Scott, 'Songs of Humpback Whales', *Science*, 13 August 1971

Priyadarshana, Tilak, et al., 2016, 'Distribution Patterns of Blue Whale (Balaenoptera musculus) and Shipping off Southern Sri Lanka', *Regional Studies in Marine Science*, January 2016

Rice, A., et al., 'Update on Frequency Decline of Northeast Pacific Blue Whale (*Balaenoptera musculus*) Calls', *PLoS One* 17, e0266469, 2022

Rothenberg, David, 2008, *Thousand Mile Song*, Basic Books

Schiffman, Richard, 'How Ocean Noise Pollution Wreaks Havoc on Marine Life', Yale E360, 31 Mar 2016, https://e360.yale.edu/features/how_ocean_noise_pollution_wreaks_havoc_on_marine_life

Srinivasan, Amia, 'What Have We Done to the Whale?', *The New Yorker*, 17 August 2020, https://www.newyorker.com/magazine/2020/08/24/what-have-we-done-to-the-whale

Leviathan, or the Sperm Whale

Anon., 'Discovery of Sound in the Sea: Sperm Whale', https://dosits.org/galleries/audio-gallery/marine-mammals/toothed-whales/sperm-whale/

Fais, A., et al., 'Sperm Whale Predator-prey Interactions Involve Chasing and Buzzing, but no Acoustic Stunning', *Scientific Reports*, 6, 28562, 24 June 2016

Hoare, Philip, 2015, review of *The Cultural Lives of Whales and Dolphins* by Hal Whitehead and Luke Rendell, *Guardian*, 10 January 2015, https://www.theguardian.com/books/2015/jan/10/cultural-lives-of-whales-and-dolphins-hal-whitehead-luke-rendell-review

Hoare, Philip, 'A Moment That Changed Me – Looking a Sperm Whale in the Eye', *Guardian*, 10 September 2015, https://www.theguardian.com/commentisfree/2015/sep/10/sperm-whale-azores-ocean-pod

Melville, Herman, 1851, *Moby Dick* (Chapter 74), Gutenberg.org

Moore, Michael J., 2022, *We Are All Whalers: The Plight of Whales and Our Responsibility*, University of Chicago Press

Nestor, James, 2016, 'A Conversation with Whales', *New York Times*, 16 April 2016, https://www.nytimes.com/interactive/2016/04/16/opinion/sunday/conversation-with-whales.html

Safina, Carl, 2020, *Becoming Wild: How Animals Learn to Be Animals*, Oneworld

Schafer, R. Murray, 1977, *The Soundscape: Our Sonic Environment and the Tuning of the World*, Knopf

Taylor, B. L., et al., *Physeter macrocephalus* (amended version of 2008 assessment), The IUCN Red List of Threatened Species, 2019

Whitehead, Hal, and Rendell, Luke, 2015, *The Cultural Lives of Whales and Dolphins*, University of Chicago Press

Blackbird

Burns, Fiona, et al., 'Abundance Decline in the Avifauna of the European Union Reveals Global Similarities in Biodiversity Change: Input Datasets & Species Results', Zenodo, 1 October 2021, https://zenodo.org/record/5544548#.Y2DarezP3Xo

Cocker, Mark, and Mabey, Richard, 2005, *Birds Britannica*, Chatto & Windus

Doolittle, Ford, 'Is Earth an Organism?', *Aeon*, 3 December 2020, https://aeon.co/essays/the-gaia-hypothesis-reimagined-by-one-of-its-key-sceptics

Lane, Nick, 2022, *Transformer: The Deep Chemistry of Life and Death*, Profile

Lawrence, D. H. (1917) 2019, 'Whistling of Birds', reprinted in *Life with a Capital L: Essays Chosen and Introduced by Geoff Dyer*, Penguin

McCartney, Paul, 1968, 'Blackbird', *The White Album*, Apple

Thomas, Edward, 'Adlestrop', 2019, *Selected Poems and Prose*, Penguin Classics

Thomas, Edward, diary entry for 11 March 1917, War Diary, The Edward Thomas Literary Estate via First World War Poetry Digital Archive, accessed 10 July 2022, http://ww1lit.nsms.ox.ac.uk/ww1lit/collections/document/1693

Safina, Carl, 2020, 'Mother Culture', *Orion Magazine*, 19 May 2020, https://orionmagazine.org/article/mother-culture/

Schnitzler, Joseph G., et al., 'Size and Shape Variations of the Bony Components of Sperm Whale Cochleae', *Scientific Reports*, 25 April 2017, DOI:10.1038/srep46734

Smyth, Richard, 2017, *A Sweet, Wild Note: What We Hear When the Birds Sing*, Elliott and Thompson

Uhrich, Alex, et al., 2020, 'How Air Sacs Power Lungs in Birds' Respiratory System', ask nature.org, https://asknature.org/strategy/respiratory-system-facilitates-efficient-gas-exchange/

Yong, Ed, 2022, *An Immense World: How Animal Senses Reveal the Hidden Realms Around Us*, Bodley Head

Owl

Calvez, Leigh, 2016, *The Hidden Lives of Owls: The Science and Spirit of Nature's Most Elusive Birds*, Sasquatch Books

Choiniere, Jonah N., et al., 'Evolution of Vision and Hearing Modalities in Theropod Dinosaurs', *Science*, 372(6542), 7 May 2021

Coombs, E. J., et al., 'Wonky Whales: The Evolution of Cranial Asymmetry in Cetaceans', *BMC Biology*, 10 July 2020

Evans Ogden, Lesley, 'The Silent Flight of Owls Explained', audobon.org, 28 July 2017, https://www.audubon.org/news/the-silent-flight-owls-explained

Graham, Robert Rule, 'The Silent Flight of Owls', *The Aeronautical Journal*, 38(286), October 1934, pp. 837–43

Ichioka, Sarah, and Pawlyn, Michael, 2021, *Flourish: Design Paradigms for Our Planetary Emergency*, Triarchy Press

Jaworski, Justin W., 'Aeroacoustics of Silent Owl Flight', *Annual Review of Fluid Mechanics*, 28 August 2019

Krumm, Bianca, 'Barn Owls Have Ageless Ears', *Proceedings of the Royal Society B*, 20 September 2017

Macintyre, Ken, 'Report on the "Owl Stone" Aboriginal Site at Red Hill, Northeast of Perth', *Anthropology from the Shed*, April 2009, https://anthropologyfromtheshed.com/project/report-owl-stone-aboriginal-site-red-hill-northeast-perth/

Mackenzie, Dana, 'The Silence of the Owls', *Knowable Magazine*,

7 April 2020, https://knowablemagazine.org/article/
    technology/2020/how-owls-fly-without-making-a-sound
'MBARI's Top 10 Deep-sea Animals', 19 September 2020, Monterey
    Bay Aquarium Research Institute, https://www.youtube.com/
    watch?v=8oOG2BGrmyA
Moubayidin, Laila, 'The Science of Symmetry', The Royal
    Society, October 2022, https://www.youtube.com/
    watch?v=K8JxMQds-PI
Norberg, R. A., 'Occurrence and Independent Evolution of Bilateral
    Ear Asymmetry in Owls and Implications on Owl Taxonomy',
    Philosophical Transactions of the Royal Society B, 31 April 1977
Pavid, Katie, 'Echolocation Gives Whales Lopsided Heads',
    nhm.ac.uk, 10 July 2020, https://www.nhm.ac.uk/discover/
    news/2020/july/echolocation-gives-whales-lopsided-heads.html
Pawlyn, Michael, 2010, 'Using Nature's Genius in
    Architecture', TED Talk, https://www.ted.com/talks/
    michael_pawlyn_using_nature_s_genius_in_architecture
Slaght, Jonathan C., 2020, *Owls of the Eastern Ice: The Quest to Find
    and Save the World's Largest Owl*, Farrar, Straus and Giroux
Smith, Lucy (producer, director), 2015–16, 'Super Powered Owls',
    BBC Two series, Natural World, https://www.bbc.co.uk/
    programmes/b054fn09
Tate, Peter, 2007, *Flights of Fancy: Birds in Myth, Legend and
    Superstition*, Random House
Wagner, Hermann, et al., 'Features of Owl Wings That Promote
    Silent Flight', *Interface Focus*, 6 February 2017
Weinberger, Eliot, 2020, *Angels and Saints*, New Directions

Nightingale

Alberge, Dalya, 'The Cello and the Nightingale: 1924
    Duet Was Faked, BBC admits', *Guardian*, 8 April 2022,
    https://www.theguardian.com/media/2022/apr/08/
    the-cello-and-the-nightingale-1924-duet-was-faked-bbc-admits
Bates, Herbert Ernest, 1936, 'Oak and Nightingale', in *Through the
    Woods*, Victor Gollancz
Birkhead, Mike, and Jones, Beth (directors), 2022, *Attenborough's*

Wonder of Song, BBC One, https://www.bbc.co.uk/programmes/
m00134jr

Brumm, Henrik, and Todt, Dietmar, 'Male–male Vocal Interactions
and the Adjustment of Song Amplitude in a Territorial Bird',
Animal Behaviour, February 2004

Falk, Dan, 'Anil Seth Finds Consciousness in Life's Push Against
Entropy', *Quanta Magazine*, 30 September 2021, https://www.
quantamagazine.org/anil-seth-finds-consciousness-in-lifes-push-
against-entropy-20210930/

Fishbein, Adam, 'How Birds Hear Birdsong', *Scientific American*, 1
May 2022

Ghosh, Amitav, 2021, *The Nutmeg's Curse: Parables for a Planet in
Crisis*, John Murray

Grange, Jeremy (producer), 2021, 'The Nightingales of Berlin',
Between the Ears, BBC Radio 3, https://www.bbc.co.uk/sounds/
play/m000wslh

Henderson, Caspar, 'A Nightingale Sang', Perspectiva Inside
Out (blog), 30 April 2019, https://perspectivainsideoutcom.
wordpress.com/2019/04/30/a-nightingale-sang/

Inspired by Iceland, 2021, 'Introducing the Icelandverse', https://
www.youtube.com/watch?v=enMwwQy_noI

Knight, Will, 'Urban Nightingales' Songs Are Illegally Loud', *New
Scientist*, 5 May 2004

Lawrence, D. H., 1917, 'Whistling of Birds', *Athenæum*, 11 April 1919,
and included in *Life with a Capital L: Essays Chosen and Introduced by
Geoff Dyer*, Penguin 2019. See also Lawrence, D. H., 1932, *Sketches of
Etruscan Places and Other Italian Essays*, Viking: 'Before Buddha or
Jesus spoke the nightingale sang, and long after the words of Jesus
and Buddha are gone into oblivion the nightingale still will sing'.

Lee, Sam, 2021, *The Nightingale: Notes on a Songbird*, Century

Liu, Wendy, 2020, *Abolish Silicon Valley: How to Liberate Technology
from Capitalism*, Repeater Books

Lyon, Pamela, 'On the Origin of Minds', *Aeon*, 21 October 2021,
https://aeon.co/essays/the-study-of-the-mind-needs-a-copernican-
shift-in-perspective

Rothenberg, David, and Erel, Korhan, 2015, *Berlin Bülbül*, Terra
Nova Music

Rothenberg, David, 2019, *Nightingales in Berlin: Searching for the Perfect Sounds*, University of Chicago Press

Seatter, Robert, 'The Cello and the Nightingale', *BBC Magazine*, 25 March 2016, https://www.bbc.co.uk/news/magazine-35861899

Smith, Harrison, 'The $13 Trillion Fantasy: Ideology and Finance in the Metaverse', Political Economy Research Centre, 24 October 2022, https://www.perc.org.uk/project_posts/the-13-trillion-fantasy-ideology-and-finance-in-the-metaverse/

Smyth, Richard, 2017, *A Sweet, Wild Note: What We Hear When the Birds Sing*, Elliot and Thompson

Tanttu, Ville (director), 2020, *Nightingales in Berlin* (film), nightingalesinberlin.com

Wiener, Anna, 'Money in the Metaverse', *The New Yorker*, 4 January 2022, https://www.newyorker.com/news/letter-from-silicon-valley/money-in-the-metaverse

Rhythm (3) – Music and Dance

Arleo, Andy, '"Fascinating Rhythm's" Fascinating Rhythm: Celebrating the Gershwins' Self-referential Song', *Imaginaires*, Presses Universitaires de Reims, 2005

Barletta, Vincent, 2020, *Rhythm: Form and Dispossession*, University of Chicago Press

Berger, Kevin, 'Rhythm's the Thing: Pianist Vijay Iyer Gives us a Master Class in the Science of Rhythm', *Nautilus*, 23 October 2014, https://nautil.us/genius-is-in-the-groove-2500/

Bruce, David, 2019, 'Extreme Math Nerd Music (An Introduction to Konnakol)', https://www.youtube.com/watch?v=OyyfLtYQcwI

Buskirk, Don (undated), 'Bulgarian Dance Rhythms, Folkdance Footnotes, https://folkdancefootnotes.org/dance/dance-information/bulgarian-dance-rhythms/

Chai, David, 'There Has Never Been a Time When This Article Didn't Exist', *Psyche*, 17 February 2021, https://psyche.co/ideas/there-has-never-been-a-time-when-this-article-didnt-exist

Gabay, Yogev, 2021, 'Chaabi. That Moroccan groove that made you go WHHAAATTTTTT', https://www.youtube.com/watch?v=xM83XVw83yk&ab_channel=YogevGabay

Gilbert, Jeremy, 2014, *Common Ground Democracy and Collectivity in an Age of Individualism*, Pluto

Gilbert, Jeremy, 'A God That Knows How to Dance', Jeremy Gilbert Writing (blog), 24 November 2020, https://jeremygilbertwriting. wordpress.com/2020/11/24/a-god-that-knows-how-to-dance/

Hamilton, Andy, et al. (eds), 2019, *The Philosophy of Rhythm: Aesthetics, Music, Poetics*, Oxford University Press

Ito, Yoshiki, et al., 'Spontaneous Beat Synchronization in Rats: Neural Dynamics and Motor Entrainment', *Science Advances*, 11 November 2022

Iyer, Vijay, 'Strength in Numbers: How Fibonacci Taught Us How to Swing', *Guardian*, 15 October 2009, https://www.theguardian. com/music/2009/oct/15/fibonacci-golden-ratio

LaMothe, Kimerer, 'The Dancing Species: How Moving Together in Time Helps Make Us Human', *Aeon*, 4 June 2019, https://aeon. co/ideas/the-dancing-species-how-moving-together-in-time-helps-make-us-human

Marsden, Rhodri, Twitter, 20 April 2021, https://twitter.com/ rhodri/status/1384421832657412097

Matacic, Catherine, 'Rhythm Might Be Hardwired in Humans', *Science*, 19 December 2016

Mehr, Samuel A., et al., 'Origins of Music in Credible Signalling', *Behavioral and Brain Sciences*, 26 August 2020

Schrofer, Jasmijn, 2015, *Tarikat* ('The Path'), Netherlands Film Academy

Shivapriya, V., and Somashekar Jois, B. R., 2018, Konnakol Duet, MadRasana Unplugged Season 03 Episode 01, https://www. youtube.com/watch?v=iurhjlBum00

Spitzer, Michael, 2021, *The Musical Human: A History of Life on Earth*, Bloomsbury

Taronga Zoo, Twitter, 31 August 2021, https://twitter.com/ tarongazoo/status/1432489666897453057

Onomatopoeia

Rousseau, Bryant, 'Which Language Uses the Most Sounds? Click 5 Times for the Answer', *New York Times*, 25 November 2016, https://www.nytimes.com/2016/11/25/world/what-in-the-world/click-languages-taa-xoon-xoo-botswana.html

How Language Began

Bertland, Alexander (undated), 'Giambattista Vico', *Internet Encyclopaedia of Philosophy*, https://iep.utm.edu/vico/

Cooperrider, Kensy, 'Hand to Mouth: If Language Began with Gestures Around a Campfire and Secret Signals on Hunts, Why Did Speech Come to Dominate Communication?', *Aeon*, 24 July 2020, https://aeon.co/essays/if-language-began-in-the-hands-why-did-it-ever-leave

Darwin, Charles, 1871, *The Descent of Man, and Selection in Relation to Sex*, http://darwin-online.org.uk/

Hampshire, Stuart, 'Vico and Language', *New York Review of Books*, 13 February 1969, https://www.nybooks.com/articles/1969/02/13/vico-and-language/

Hauser, Mark, et al., 'The Mystery of Language Evolution', *Frontiers in Psychology*, 7 May 2014

Joordens, J., et al., '*Homo erectus* at Trinil on Java Used Shells for Tool Production and Engraving', *Nature*, 518, 3 December 2014, pp. 228–31

Kersken, Verena, et al., 'A Gestural Repertoire of 1- to 2-Year-Old Human Children', *Animal Cognition*, 22, 8 September 2018

Leland, Andrew, 'DeafBlind Communities May Be Creating a New Language of Touch', *The New Yorker*, 12 May 2022, https://www.newyorker.com/culture/annals-of-inquiry/deafblind-communities-may-be-creating-a-new-language-of-touch

McCarthy, Cormac, 'The Kekulé Problem: Where Did Language Come From?', *Nautilus*, 17 April 2017, https://nautil.us/the-kekul-problem-236574/

Mehr, Samuel, et al., 'Origins of Music in Credible Signaling', *Behavioral and Brain Sciences*, 26 August 2020

Newman, Rob, 'On Song', episode 4 of Newman On Air, BBC Radio 4, 16 November 2022, https://www.bbc.co.uk/sounds/play/m001f5ld

Niekus, Marcel J. L., 'Middle Paleolithic Complex Technology and a

Neandertal Tar-backed Tool from the Dutch North Sea', *PNAS*, 21 October 2019

Robson, David, 'The Origins of Language Discovered in Music, Mime and Mimicry', *New Scientist*, 1 May 2019

Roy, Deb, 2011, 'The Birth of a Word', TED Talk, https://www.ted.com/talks/deb_roy_the_birth_of_a_word

Safina, Carl, 2020, *Becoming Wild: How Animals Learn to Be Animals*, Oneworld

Wragg Sykes, Rebecca, 2020, *Kindred: Neanderthal Life, Love, Death and Art*, Bloomsbury

The Magic Flute

Akhtar, Navid, 2021, 'An Introduction to Sufi Music', https://sites.barbican.org.uk/sufimusic/

Bayaka Pygmies, recorded by Louis Sarno, 'Flute in Forest', in *Song from The Forest,* www.songfromtheforest.com

Bellando, Nick, and Deschênes, Bruno, 2020, 'The Role of Tone-colour in Japanese Shakuhachi Music', *Ethnomusicology Review*, 22, 2020

Boyden, Ian, 2020, *A Forest of Names: 108 Meditations*, Wesleyan University Press

Chase, Claire, 2013–2036, Density 2036, https://www.density2036.org/

Cox, Trevor (undated), 'Sonic Wonders of the World', sonicwonders.org, http://www.sonicwonders.org/gong-rocks-serengeti-national-park/

Cuadros, Alex, 'Songs from Sinjar: How ISIS Is Hastening the End of the Yezidis' Ancient Oral Tradition', *Lapham's Quarterly: Music*, 2017, https://www.laphamsquarterly.org/music/songs-sinjar

Ehrenreich, Barbara, 'The Humanoid Stain', *The Baffler*, November 2019, https://thebaffler.com/salvos/the-humanoid-stain-ehrenreich

Freedman, Adrian, 2021, personal communication

Garner, Alan, 2021, *Treacle Walker*, 4th Estate

Kornei, Katherine, 'Hear the Sound of a Seashell Horn Found in an

Ancient French Cave', *New York Times*, 10 February 2021, https://www.nytimes.com/2021/02/10/science/conch-shell-horn.html

Krause, Bernie, 2012, *The Great Animal Orchestra: Finding the Origins of Music in the World's Wild Places*, Little, Brown and Company

Lee, Riley (undated), 'About the Shakuhachi', http://rileylee.net/about-the-shakuhachi/

Potengowski, Anna Friederike, 2017, *The Edge of Time: Paleolithic Bone Flutes from France & Germany*, Delphian Records

Quignard, Pascal, 2016, *The Hatred of Music*, Yale University Press

Rainio, Riitta, et al., 2021, 'Prehistoric Pendants as Instigators of Sound and Body Movements: A Traceological Case Study from Northeast Europe, *c.* 8200 cal. BP', *Cambridge Archaeological Journal*, 26 May 2021, https://doi.org/10.1017/S0959774321000275

Ross, Alex, 'Claire Chase Taps the Primal Power of the Flute', *The New Yorker*, 3 January 2022, https://www.newyorker.com/magazine/2022/01/03/claire-chase-taps-the-primal-power-of-the-flute

Rūmī, trans. Coleman Barks, 2004, 'Flutes for Dancing', *The Essential Rūmī*, HarperCollins

Severini, Giuseppe, 2018, 'Sounds of Nature', https://www.youtube.com/watch?v=CqAmkHXgJ_0

Stutzmann, Nathalie, et al., '5 Minutes That Will Make You Love the Flute', *New York Times*, 6 January 2021, https://www.nytimes.com/2021/01/06/arts/music/five-minutes-classical-music-flute.html

Sword, Harry, 2021, *Monolithic Undertow: In Search of Sonic Oblivion*, White Rabbit, https://mathewlyons.co.uk/2021/05/13/the-quietus-monolithic-undertow-by-harry-sword/

The Nature of Music

Cage, John, and Crumb, George, 1975, short written notes in response to a request from Edition Peters to a number of composers for their thoughts on music. A correspondent at the publisher suggested to the author that Cage had in mind Henry David Thoreau, who said, 'All sound is nearly akin to Silence; it is a bubble on her surface which straightway bursts, an emblem of the strength and prolificness of the undercurrent.' https://twitter.com/EditionPetersUS/status/1235592061115457536

King, Christopher C., 2018, *Lament from Epirus: An Odyssey into Europe's Oldest Surviving Folk Music*, W. W. Norton

Kubik, Gerhard, 1979, 'Pattern Perception and Recognition in African Music', in John Blacking and Joann W. Kealiinohomoko (eds), *The Performing Arts*, De Gruyter Mouton

Lewis, Jerome, 2013, 'A Cross-Cultural Perspective on the Significance of Music and Dance to Culture and Society: Insight from BaYaka Pygmies', in M. Arbib (ed.), *Language, Music, and the Brain*, MIT Press

Margulis, Elizabeth Hellmuth, 2019, *The Psychology of Music: A Very Short Introduction*, Oxford University Press

Neely, Adam, 2018, 'How to explain music to Aliens?', https://www.youtube.com/watch?v=cdasn27lbgY

Nettl, Bruno, 2000, 'An Ethnomusicologist Contemplates Universals in Musical Sound and Musical Culture', in N. Wallin et al. (eds), *The Origin of Music*, MIT Press

Obert, Michael, 2013, *Song from the Forest* (documentary film), http://songfromtheforest.com/

Patel, Aniruddh D., 2010, *Music, Language, and the Brain*, Oxford University Press

Peretz, Isabelle, and Zatorre, Robert, 'Brain Organization for Music Processing', *Annual Review of Psychology*, 4 February 2005

Rilke, Rainer Maria, 1918, 'An die Musik', with a commentary by Scott Horton, *Harper's Magazine*, 23 August 2009, https://harpers.org/2009/08/rilke-to-music/

Warshall, Peter, 1999, 'Two Billion Years of Animal Sounds' (lecture), Allen Ginsberg Library and Naropa University Archives, http://archives.naropa.edu/digital/collection/p16621coll1/id/1347

Williamson, Victoria, 2014, *You Are the Music: How Music Reveals What It Means to Be Human*, Icon Books

Xenakis, Iannis, 1971, 'Towards a Metamusic', in *Formalized Music: Thought and Mathematics in Composition*, Pendragon Press

Zimmer, Carl, 2021, *Life's Edge: The Search for What It Means to Be Alive*, Picador. He quotes the philosopher Carol Cleland: 'Definitions are not the proper tools for answering the scientific question "What is life?"'

Harmony

Ackerman, Diane, 1990, *A Natural History of the Senses*, Random House

Charpentier, Marc-Antoine, c.1692, *Les Règles de composition*

Collier, Jacob, 'The Music That Got Me Through 2020', *Jacob Collier's Music Room*, BBC Radio 3, 28 December 2020, https://www.bbc.co.uk/programmes/m000qlkh

Goodall, Howard, 2013, *The Story of Music*, Vintage

Hume, David, 1739, *A Treatise of Human Nature*, 3.3.1.7, SBN 575-6 https://davidhume.org/texts/t/3/3/1

Mathieu, W. A., 1997, *Harmonic Experience*, Inner Traditions. Mathieu writes: 'By resonance I mean those specially reinforcing combinations of tones that in their mutual resounding – their perfect in-tune-ness – evaporate the boundary between music and musician.'

Neely, Adam, 'Music Theory and White Supremacy', 2020, https://www.youtube.com/watch?v=Kr3quGh7pJA&t=115s

Ockelford, Adam, 2017, *Comparing Notes: How We Make Sense of Music*, Profile

Tepfer, Dan, 'Rhythm/Pitch Duality: Hear Rhythm Become Pitch Before Your Ears', (blog post), 13 December 2012, https://dantepfer.com/blog/?p=277

Strange Musical Instruments

Howes, Anton, 2022, 'Age of Invention: Why Wasn't the Steam Engine Invented Earlier? Part II' (substack), https://antonhowes.substack.com/ citing Salomon de Caus, *Les Raisons des forces mouvantes* (Jan Morton, 1615), pp. 19–21

Sad Songs

Alter, Robert, 2009, *The Book of Psalms: A Translation with Commentary*, W. W. Norton

Bekoff, Marc, 2010, *The Emotional Lives of Animals: A Leading Scientist Explores Animal Joy, Sorrow, and Empathy and Why They Matter*, New World Library

Berger, John, 2016, *Confabulations*, Penguin

Bruce, David, 2021, 'Which Instrument is the SADDEST?', https://www.youtube.com/watch?v=b-uild3eyRY

Cave, Nick, 2019, The Red Hand Files (blog), https://www.theredhandfiles.com/

Chaudhuri, Amit, 2021, *Finding the Raga: An Improvisation on Indian Music*, Faber & Faber. 'The natural third has a buoyancy that corresponds to the renewal the rains offer: the washed leaves and hair; the burgeoning of, and movement in, the environment. The flat third, following immediately, introduces reflectiveness.' https://www.theguardian.com/books/2021/apr/29/finding-the-raga-by-amit-chaudhuri-a-passion-for-indian-music

da Fonseca-Wollheim, Corinna, 'Moving on', Beginner's Ear (blog), 10 July 2021, https://www.beginnersear.com/musings/moving-on

Douglass, Frederick, (1845) (1999), *Narrative of the Life of Frederick Douglass*, Oxford University Press

Feld, Steven, 'Wept Thoughts: The Voicing of Kaluli Memories', *Oral Tradition*, May 1990, https://www.researchgate.net/publication/237596107_Wept_Thoughts_The_Voicing_of_Kaluli_rie

Fine, Sarah, 'Humanity at Night', *Aeon*, 16 November 1990, https://aeon.co/essays/in-times-of-crisis-the-arts-are-weapons-for-the-soul

Haysom, Simone, 'Fortress Conservation', review of *Security and Conservation: The Politics of the Illegal Wildlife Trade* by Rosaleen Duffy, *London Review of Books*, 1 December 2022, https://www.lrb.co.uk/the-paper/v44/n23/simone-haysom/fortress-conservation

Huron, David, 'Why Is Sad Music Pleasurable? A Possible Role for Prolactin', *Musicae Scientiae*, 15(2), 2011

Huron, David, and Vuoskoski, Jonna K., 'On the Enjoyment of Sad Music: Pleasurable Compassion Theory and the Role of Trait Empathy', *Frontiers in Psychology*, 20 May 2020

Ignatieff, Michael, 2021, *On Consolation: Finding Solace in Dark Times*, Picador

Kallison, David, 2018, 'The Depression Issue', *The Sound and the Story* (podcast)

Kamieńska, Anna (undated), 'Industrious Amazement: A Notebook', *Poetry* Magazine, https://www. poetryfoundation.org/poetrymagazine/articles/69655/ industrious-amazement-a-notebook

King, Barbara J., 2013, *How Elephants Grieve*, University of Chicago Press

King, Barbara J., 2019, 'Grief and Love in the Animal Kingdom', TED Talk, https://www.ted.com/talks/ barbara_j_king_grief_and_love_in_the_animal_kingdom

McLeish, Tom, 2020, 'Thought for the Day', BBC Radio 4, 8 December 2020, https://www.bbc.co.uk/programmes/p090cqs6 https://www.bbc.co.uk/sounds/play/p0b8sd5k

Oishi, S., and Westgate, E. C., 'A Psychologically Rich Life: Beyond Happiness and Meaning', *Psychological Review*, 129(4), 2022

Pinnock, Trevor, 2017, 'Why Dido's Lament Breaks Our Heart Every Single Time', Classic FM, 24 March 2017, https://www.classicfm.com/composers/purcell/guides/ trevor-pinnock-didos-lament/

Service, Tom, 'Klezmer', *The Listening Service*, BBC Radio 3, 7 November 2021, https://www.bbc.co.uk/sounds/play/m0011clx

Shostakovich, Dmitri, 1965, 'The Power of Music', *Music Journal*, University of Washington Press

Spitzer, Michael, 'Music and Sex', *Aeon*, 18 October 2021, https:// aeon.co/essays/can-music-give-you-an-orgasm-the-short-answer- is-yes

Thomas, Lewis, 1983, *Late Night Thoughts on Listening to Mahler's Ninth Symphony*, Viking

Van den Tol, A. J. M., and Edwards, J., 'Exploring a Rationale for

Choosing to Listen to Sad Music When Feeling Sad', *Psychology of Music*, 21 December 2011

Young, Emma, 'A Sad Kind of Happiness: The Role of Mixed Emotions in Our Lives', *Research Digest*, The British Psychological Society, 25 November 2021

Bashō

Parkes, Graham and Loughnane, Adam, 'Japanese Aesthetics', *Stanford Encyclopedia of Philosophy* (Winter 2018 edition), Edward N. Zalta (ed.) https://plato.stanford.edu/entries/japanese-aesthetics/

Robbins, Jeff and Shoko, Sakata, 2018, 'Bashō for Humanity', https://www.basho4humanity.com/

Thoreau, Henry David, 1854, *Walden*, Chapter 2: 'Where I Lived, and What I Lived For', in *Walden and Other Writings*, Modern Library, 1992

Visible Sound

Cox, Trevor, 'Breaking Glass with Sound', Salford Acoustics (blog), https://salfordacoustics.co.uk/how-to-breaking-glass-with-sound

'Does Sound, Like Light, Have a Maximum Speed?', *The Economist*, 17 October 2020, https://www.economist.com/science-and-technology/2020/10/17/does-sound-like-light-have-a-maximum-speed

Feynman, Richard, 1983, 'Fun to Imagine 8', 'Seeing Things', uploaded 2009, https://www.youtube.com/watch?v=iqQQXTMih1A

Fiorella, Giancarlo, 'How to Maintain Mental Hygiene as an Open Source Researcher', Bellingcat, 23 November 2022, https://www.bellingcat.com/resources/2022/11/23/how-to-maintain-mental-hygiene-as-an-open-source-researcher/

Gioia, Ted, 2021, 'The Man Who Put Out Fires with Music', The Honest Broker (blog), 7 June 2021, https://tedgioia.substack.com/p/the-man-who-put-out-fires-with-music

Harvard Natural Sciences Lecture Demonstrations (undated),
     https://sciencedemonstrations.fas.harvard.edu/presentations/
     chladni-plates
Patrikarakos, David, 2022, 'The Demolition of Kharkiv',
     UnHerd, 11 May 2022, https://unherd.com/2022/05/
     the-demolition-of-kharkiv/

Plato's Cave

Anon, Jean Cocteau's *Orphée* (1950), Radio Transmissions,
     Rorschach Audio (blog), 22 May 2013, https://rorschachaudio.
     com/2013/05/22/cocteau-orpheus-transmissions/
Dyer, Geoff, 2012, *Zona: A Book About a Film About a Journey to a
     Room*, Canongate
Goldmanis, Māris, 'Explaining the "Mystery" of Numbers Stations',
     War on the Rocks (blog), 24 May 2018, https://warontherocks.
     com/2018/05/explaining-the-mystery-of-numbers-stations/
Khlebnikov, Velimir, 1921, 'The Radio of the Future', Museum of
     Imaginary Musical Instruments, http://imaginaryinstruments.
     org/the-radio-of-the-future/
Knight, Sam, 'Can the BBC Survive the British
     Government?', *The New Yorker*, 18 April 2022, https://
     www.newyorker.com/magazine/2022/04/18/
     can-the-bbc-survive-the-british-government
Latour, Bruno, 2017, *Facing Gaia: Eight Lectures on the New Climatic
     Regime*, Wiley
Levi, Primo, 1975, *Il sistema periodico*, Einaudi
Ritchie, Hannah, and Roser, Max, 2022, 'CO2 Emissions', *Our World
     in Data*, https://ourworldindata.org/co2-emissions
Smith, Stefan, 'The Edge of Perception: Sound in Tarkovsky's
     Stalker', *The Soundtrack*, November 2007

Earworms

Hellmuth Margulis, Elizabeth, 2013, *On Repeat: How Music Plays the
     Mind*, Oxford University Press

Sacks, Oliver, 'Musical Ears', *London Review of Books*, 3 May 1984, https://www.lrb.co.uk/the-paper/v06/n08/oliver-sacks/musical-ears

Sacks, Oliver, 2007, *Musicophilia: Tales of Music and the Brain*, Knopf

Twain, Mark, (1876) (2012), 'A Literary Nightmare'/'Punch, Brothers, Punch', in *The Complete Short Stories*, Everyman

White, E. B., 1933, 'The Supremacy of Uruguay', *The New Yorker*, 25 November 1933, https://archives.newyorker.com/newyorker/1933-11-25/flipbook/018/

Noise Pollution

'Airgun', Discovery of Sound in the Sea, University of Rhode Island, https://dosits.org/glossary/airgun/

Barber, Jesse, et al., 'An Experimental Investigation into the Effects of Traffic Noise on Distributions of Birds: Avoiding the Phantom Road', *Proceedings of the Royal Society B*, 22 December 2013

Battson, Ginny, 'Anthrophonalgia – a Plague', Seasonlight (blog), 9 April 2021, https://seasonalight.com/2021/04/09/anthrophonalgia-a-plague/

Berdik, Chris, 2020, 'The Fight to Curb a Health Scourge in India: Noise Pollution', *Undark*, 26 March 2020, https://pulitzercenter.org/stories/fight-curb-health-scourge-india-noise-pollution

Berthold, Daniel, 'Aldo Leopold: In Search of a Poetic Science', *Human Ecology Review*, 11(3), 2004, pp. 205–14

Cave, David, et al., 'Long Slide Looms for World Population, With Sweeping Ramifications', *New York Times*, 22 May 2021, https://www.nytimes.com/2021/05/22/world/global-population-shrinking.html

Chepesiuk, R., 'Decibel Hell: The Effects of Living in a Noisy World', *Environmental Health Perspectives*, 1 January 2005

Cuff, Madeleine, 'Hedgehogs Use Coronavirus Lockdown to Indulge in "Noisy Lovemaking" and Experts Predict Baby Boom', inews.co.uk, 22 May 2020, https://inews.co.uk/news/hedgehogs-coronavirus-lockdown-noisy-lovemaking-baby-boom-430507

Derryberry, Elizabeth, et al., 'Singing in a Silent Spring: Birds

Respond to a Half-century Soundscape Reversion During the
COVID-19 Shutdown', *Science*, 24 September 2020

Ghosh, Amitav, 2021, *The Nutmeg's Curse: Parables for a Planet in
Crisis*, John Murray

Haskell, David George, 2022, *Sounds Wild and Broken: Sonic Marvels,
Evolution's Creativity and the Crisis of Sensory Extinction*, Viking

Hempton, Gordon (undated), 'One Square Inch of Silence', https://
onesquareinch.org/

Hendy, David, 2013, *Noise: A Human History of Sound and Listening*,
Profile

Kearney, Martha, 'Reignite', BBC Radio 4, 28 March 2021, https://
www.bbc.co.uk/programmes/m000tm9r

Koshy, Yohann, 'The Last Humanist: How Paul Gilroy Became the
Most Vital Guide to Our Age of Crisis', *Guardian*, 5 August 2021,
https://www.theguardian.com/news/2021/aug/05/paul-gilroy-
britain-scholar-race-humanism-vital-guide-age-of-crisis

Kraus, Nina, 2021, *Of Sound Mind: How Our Brain Constructs a
Meaningful Sonic World*, MIT Press

Leopold, Aldo, (1949) 1987, 'Song of the Gavilan', in *A Sand County
Almanac: And Sketches Here and There*, Oxford. See also 'A Sense
of Place', aldoleopold.org, 23 December 2017, https://www.
aldoleopold.org/post/a-sense-of-place/

Monkhouse, Joseph, 2021, 'Recreating the Lost Soundscape of Iron
Age Somerset', https://www.youtube.com/watch?v=8OQxzf
vau8U. See also Yeo, Sophie, 2020, 'What Did Somerset Sound
Like 2,000 Years Ago?', *Inkcap* (substack), https://inkcap.
substack.com/p/what-did-somerset-sound-like-2000

Monkhouse, Joseph, et al., 2022, 'Six Thousand Years of Forests',
Inkap (substack), https://www.inkcapjournal.co.uk/six-
thousand-years-of-forests/ https://www.inkcapjournal.co.uk/
six-thousand-years-of-forests/

Owen, David, 2019, 'Is Noise Pollution the Next Big Public Health
Crisis?', *The New Yorker*, 6 May 2019, https://www.newyorker.
com/magazine/2019/05/13/is-noise-pollution-the-next-big-public-
health-crisis

Possible (campaign group), 2021, 'Car Free Cities', https://www.
wearepossible.org/carfreecities

Radford, Andy, 'Silencing with Noise', Costing the Earth, BBC Radio 4, 5 May 2020, https://www.bbc.co.uk/sounds/play/mooohtxl

Renkl, Margaret, 'The First Thing We Do, Let's Kill All the Leaf Blowers', *New York Times*, 25 October 2021, https://www.nytimes.com/2021/10/25/opinion/leaf-blowers-california-emissions.html

Rincon, Paul, 'Climate Change: Carbon Emissions Show Rapid Rebound after Covid Dip', BBC News, 4 November 2021, https://www.bbc.co.uk/news/science-environment-59148520

Rolland, Rosalind, et al., 'Evidence that Ship Noise Increases Stress in Right Whales', *Proceedings of the Royal Society B*, 8 February 2012

Rosling, Hans, et al., 2018, *Factfulness: Ten Reasons We're Wrong About the World – and Why Things Are Better Than You Think*, Flatiron Books

Shannon, Graeme, 'How Noise Pollution Is Changing Animal Behaviour', *The Conversation*, 17 December 2015, https://theconversation.com/how-noise-pollution-is-changing-animal-behaviour-52339

World Health Organisation (multiple authors), 'Environmental Noise Guidelines of the European Region', https://www.euro.who.int/__data/assets/pdf_file/0009/383922/noise-guidelines-exec-sum-eng.pdf

The Sounds of Climate Change

Anderson, Craig, 'Heat and Violence', *Current Directions in Psychological Science*, February 2001

Brunt, Kieran, et al., 'The Rising Sea Symphony', Between the Ears, BBC Radio 3, 18 October 2020, https://www.bbc.co.uk/sounds/play/mooonkzp

Burtner, Matthew, 2019, *Glacier Music*, Ravello Records

Gibbs, Peter, 'Acoustic Ecology', Costing the Earth, BBC Radio 4, 1 March 2016, https://www.bbc.co.uk/sounds/play/b071tgby

Hayhoe, Katharine, 2021, *Saving Us: A Climate Scientist's Case for Hope and Healing in a Divided World*, Atria/One Signal Publishers

Hugonnet, R., et al., 'Accelerated Global Glacier Mass Loss in the Early Twenty-first Century', *Nature*, 592, 28 April 2021, pp. 726–31

IPCC, 2018: 'Global Warming of 1.5°C. An IPCC Special Report on the Impacts of Global Warming of 1.5°C Above Pre-industrial Levels and Related Global Greenhouse Gas Emission Pathways, in the Context of Strengthening the Global Response to the Threat of Climate Change, Sustainable Development, and Efforts to Eradicate Poverty', World Meteorological Organization, Geneva, https://www.ipcc.ch/sr15/

Jamie, Kathleen, 2021, 'What the Clyde Said', https://www.scottishpoetrylibrary.org.uk/poem/what-the-clyde-said-after-cop26/

Jamie, Kathleen, 'Stay Alive! Stay Alive!', *London Review of Books*, 18 August 2022, https://www.lrb.co.uk/the-paper/v44/n16/kathleen-jamie/diary

Krause, Bernie, 2017, 'Biophony', *Anthropocene Magazine*, https://www.anthropocenemagazine.org/2017/08/biophony/

Macfarlane, Robert, 2019, *Underland: A Deep Time Journey*, Penguin Random House

Orlowski, Jeff, et al., 2012, *Chasing Ice* (documentary film), excerpt on https://www.youtube.com/watch?v=hC3VTgIP0GU

Watts, Jonathan, 'The Sound of Icebergs Melting: My Journey into the Antarctic', *Guardian*, 9 April 2020, https://www.theguardian.com/world/ng-interactive/2020/apr/09/sound-of-icebergs-melting-journey-into-antarctic-jonathan-watts-greenpeace

Watts, Jonathan, and Kommenda, Niko, 'Speed at Which World's Glaciers Are Melting Has Doubled in 20 Years', *Guardian*, 28 April 2021, https://www.theguardian.com/environment/2021/apr/28/speed-at-which-worlds-glaciers-are-melting-has-doubled-in-20-years

Yaffa, Joshua, 2022, 'The Great Siberian Thaw', *The New Yorker*, 10 January 2022, https://www.newyorker.com/magazine/2022/01/17/the-great-siberian-thaw

Xu, Chi, et al., 'Future of the Human Climate Niche', *PNAS*, 27 October 2019

Hell

Black, Jeremy, and Green, Anthony, 1992, *Gods, Demons and Symbols of Ancient Mesopotamia: An Illustrated Dictionary*, University of Texas Press

Ciabattoni, Francesco, 2019, 'Musical Instruments in Dante's *Commedia*: A Visual and Acoustic Journey', The Digital Dante, https://digitaldante.columbia.edu/sound/ciabattoni-instruments/

Dix, Otto, 1924 and 1932, *Der Krieg* (The War)

Evans, Paul, and Watson, Chris, 'The Island of Secrets', BBC Radio 4, 25 March 2009, https://www.bbc.co.uk/programmes/b00j7528

Flannery, Tim, 1998, *Throwim Way Leg*, Text Publishing

Homer, trans. Emily Wilson, 2018, *The Odyssey*, W. W. Norton

Kaminsky, Ilya, 2021, 'I See a Silence', *Afterness*, Art Angel

Macfarlane, Robert, with Somogyi, Arnie, and Wilson, Jane and Louise, 2012, *Untrue Island*, Commissions East and National Trust

Moynihan, Thomas, and Sandberg, Anders, 'Drugs, Robots and the Pursuit of Pleasure – Why Experts Are Worried about Ais Becoming Addicts', *The Conversation*, 14 September 2021, https://theconversation.com/drugs-robots-and-the-pursuit-of-pleasure-why-experts-are-worried-about-ais-becoming-addicts-163376

Pullman, Philip, 'The Sound and the Story: Exploring the World of *Paradise Lost*', *Public Domain Review*, 11 December 2019, https://publicdomainreview.org/essay/the-sound-and-the-story-exploring-the-world-of-paradise-lost

Spalink, James, 2014, 'Hieronymus Bosch Butt Music', melody based on a transcription by Amelia Hamrick, https://www.youtube.com/watch?v=OnrICy3Bc2U

Steggle, Matthew, 2001, '*Paradise Lost* and the Acoustics of Hell', *Early Modern Literary Studies*, 7(1), special issue 8

Healing with Music

American Music Therapy Association, 'What Is Music Therapy?', 'clinical and evidence-based use of music interventions to accomplish individualized goals within a therapeutic relationship', https://www.musictherapy.org/about/musictherapy/, accessed 2 September 2021

Burkeman, Oliver, 2021, *Four Thousand Weeks: Time and How to Use It*, Bodley Head

Cypess, Rebecca, 'Giovanni Battista Della Porta's Experiments with Musical Instruments', *Journal of Musicological Research*, 35(3), pp. 159–75, 26 May 2016

Ficino, Marsilio, (1496) 2010, *'All Things Natural': Ficino on Plato's Timaeus*, Shepheard-Walwyn

Garrow, Duncan, and Wilkin, Neil, 2022, *The World of Stonehenge* (catalogue), British Museum

Gergis, Joëlle, et al., 2021, 'We're Not About to Back Down: How Climate Experts Hold Hope Despite the IPCC Report', *Guardian*, 10 August 2021, https://www.theguardian.com/commentisfree/2021/aug/10/were-not-about-to-back-down-how-climate-experts-hold-hope-despite-the-ipcc-report

Gioia, Ted, 2006, *Healing Songs*, Duke University Press

Grünberg, Judith M., et al., 2013, 'Analyses of Mesolithic Grave Goods from Upright Seated Individuals in Central Germany', in *Mesolithic Burials – Rites, Symbols and Social Organisation of Early Postglacial Communities*, Landesamt für Denkmalpflege und Archäologie Sachsen-Anhalt

Katz, Richard, 1982, *Boiling Energy: Community Healing Among the Kalahari !Kung*, Harvard University Press

Kleisiaris, Christos F., et al., 'Health Care Practices in Ancient Greece: The Hippocratic Ideal', *Journal of Medical Ethics and History of Medicine*, 7(6), March 2014

Natural Healing Society (undated), 'Solfeggio Frequencies', Natural Healing Society, https://www.naturehealingsociety.com/articles/solfeggio/

Puhan, M. A., et al., 'Didgeridoo Playing as Alternative Treatment for Obstructive Sleep Apnoea Syndrome: Randomised Controlled Trial', *BMJ*, 4 February 2006

Sacks, Oliver, 2007, *Musicophilia: Tales of Music and the Brain*, Picador

Service, Tom, 'Is Music Good for You?', *The Listening Service*, BBC Radio 3, 9 May 2021, https://www.bbc.co.uk/sounds/play/mooovwpx

Sloboda, John, 2005, *Exploring the Musical Mind: Cognition, Emotion, Ability, Function*, Oxford University Press

Sloboda, John, 'The Ear of the Beholder', *Nature*, 3 July 2008

Tomatis, Alfred, 1991, *The Conscious Ear: My Life of Transformation Through Listening*, Station Hill Press

Truer, David, 2021, 'A Sadness I Can't Carry: The Story of the Drum', *New York Times*, 31 August 2021, https://www.nytimes.com/2021/08/31/magazine/ojibwe-big-drum.html

'Who Was this Mysterious Ballerina from the Viral Swan Lake Video?', CBC Radio, 16 November 2021, https://www.cbc.ca/radio/thecurrent/the-current-for-nov-16-2020-1.5803389/who-was-this-mysterious-ballerina-from-the-viral-swan-lake-video-1.5803747

Williamson, Victoria, 2014, *You Are the Music: How Music Reveals What It Means to Be Human*, Icon Books

Healing with Sound

Alvarsson, J. J., et al., 'Stress Recovery During Exposure to Nature Sound and Environmental Noise', *International Journal of Environmental Research and Public Health*, 7(3), 20 February 2010, pp. 1036–46

Bates, Victoria, 2021, *Making Noise in the Modern Hospital*, Cambridge University Press

Bates, Victoria, 2022, 'How the Noises of a Hospital Can Become a Healing Soundscape', *Aeon*, 8 February 2022, https://psyche.co/ideas/how-the-noises-of-a-hospital-can-become-a-healing-soundscape

Batty, David, 'Bird and Birdsong Encounters Improve Mental Health', *Guardian*, 27 October 2022, citing Hammoud, Ryan, et al., 'Smartphone-based Ecological Momentary Assessment Reveals Mental Health Benefits of Birdlife', *Nature*, 27 October 2022

Gould van Praag, C., et al., 'Mind-wandering and Alterations
    to Default Mode Network Connectivity When Listening to
    Naturalistic Versus Artificial Sounds', *Scientific Reports* 7, 45273,
    27 March 2017
Institute of Cancer Research, 'World First Treatment with
    "Acoustic Cluster Therapy" to Improve Chemotherapy
    Delivery', press release, 18 December 2019, https://www.icr.
    ac.uk/news-archive/world-first-treatment-with-acoustic-cluster-
    therapy-to-improve-chemotherapy-delivery
Jones, Lucy, 2020, *Losing Eden: Why Our Minds Need the Wild*, Allen
    Lane
O'Reilly, Sally, 'The Devil's Chord or a Tap on the Shoulder?
    Recomposing the Soundscape of the Intensive Care Unit',
    Cabinet, 9 July 2020, https://www.cabinetmagazine.org/kiosk/
    oreilly_sally_9_july_2020.php
Sanguinetti, Joseph L., et al., 'Transcranial Focused Ultrasound
    to the Right Prefrontal Cortex Improves Mood and Alters
    Functional Connectivity in Humans', *Frontiers in Human
    Neuroscience*, 28 February 2020
Shrivastava, Shamit, 'A Sound Future for Noninvasive Therapies',
    19 August 2017, https://medium.com/@Shamits/
    a-sound-future-for-noninvasive-therapies-ac487ea03977
Wood, Emma, et al., 'Not All Green Space Is Created Equal:
    Biodiversity Predicts Psychological Restorative Benefits From
    Urban Green Space', *Frontiers in Psychology*, 27 November 2018

Bells

Ap Myrddin, Llywelyn, 'Russian Bells', BBC Radio 4, 9 October
    2017, https://www.bbc.co.uk/sounds/play/b0978ndz
Australian Bell, 'New Technologies', http://www.ausbell.com.au/
    new_tech.html
Batuman, Elif, 'The Bells: How Harvard Helped Preserve a Russian
    Legacy', *The New Yorker*, 27 April 2009, https://www.newyorker.
    com/magazine/2009/04/27/the-bells-6
Brand, Stewart, 2000, *The Clock of the Long Now: Time and
    Responsibility*, Basic Books

Braun, Martin, 'Bell Tuning in Ancient China: A Six-tone Scale in a 12-tone System Based on Fifths and Thirds', *Neuroscience of Music*, 16 June 2003

Bruce, David, 2021, 'Why Composers Love Bells', https://www.youtube.com/watch?v=Ii3BwiU7leg

Conen, Hermann, 'White Light', essay trans. into English by Eileen Walliser-Schwarzbart (found in the liner notes of the ECM release of Arvo Pärt's *Alina*)

Dillard, Annie, 1974, *Pilgrim at Tinker Creek*, Harper's Magazine Press

Enfield, Lizzie, 2022, 'Dunwich: The British Town Lost to the Sea', BBC Travel, 27 February 2022, https://www.bbc.com/travel/article/20220227-dunwich-the-british-town-lost-to-the-sea

Eno, Brian, 2003, 'About Bells', liner notes for the album *January 07003: Bell Studies for The Clock of The Long Now*, Opal Music

Eno, Brian (undated), 'The Big Here and Long Now' (blog post), https://longnow.org/essays/big-here-long-now/

Finer, Jem, with Artangel, 1999, longplayer.org. See also 'The Longplayer Conversation', 2017, https://longplayer.org/conversations/the-longplayer-conversation-2017/

Gioia, Ted, 2006, *Healing Songs*, Duke University Press

Graça da Silva, Sara, and Tehrani, Jamshid J., 2016, 'Comparative Phylogenetic Analyses Uncover the Ancient Roots of Indo-European Folktales', *Royal Society Open Science* 3:150645150645, 1 January 2016, https://doi.org/10.1098/rsos.150645

Gray, John, 2011, *The Immortalisation Commission: The Strange Quest to Cheat Death*, Allen Lane

Harvey, Jonathan, 2005, 'Cut and Splice', BBC Radio 3, http://www.bbc.co.uk/radio3/cutandsplice/mortuos.shtml

Jayarava, Dharmacari, undated, 'Cantus in Memory of Benjamin Britten by Arvo Pärt', online essay at http://jayarava.org/cantus.html

Jerram, Luke, 2019, 'Extinction Bell', https://www.lukejerram.com/extinction-bell/

Král, Petr, 'Tarkovsky, or the Burning House', translated from the Czech by Kevin Windle. Originally published in *Svedectvi* XXIII, 91, 1990, pp. 258–68. Reproduced in *Screening the Past*, 1 March 2001

Landers, Jackson, 2017, 'A Rare Collection of Bronze Age Chinese Bells Tells a Story of Ancient Innovation', *Smithsonian Magazine*, 5 October 2017, https://www.smithsonianmag.com/smithsonian-institution/bronze-age-chinese-bells-tells-story-ancient-innovation-180964459/

Lienhard, John H., *Engines of Our Ingenuity*, no. 1676: 'Ancient Chinese Bells', https://www.uh.edu/engines/epi1676.htm

McLachlan, Neil, and Nigjeh, Behzad, 'The Design of Bells with Harmonic Overtones', *The Journal of the Acoustical Society of America*, 3 July 2003

Ramm, Benjamin, 2021, 'Cosmism: Russia's Religion for the Rocket Age', BBC Future, 20 April 2021, https://www.bbc.com/future/article/20210420-cosmism-russias-religion-for-the-rocket-age

Roberts, Adams, 2021, *Middlemarch: Epigraphs and Mirrors*, Open Book Publishers

Sherman, Anna, 2019, *The Bells of Old Tokyo: Meditations on Time and a City*, Pan MacMillan

Spitzer, Michael, 2021, *The Musical Human: A History of Life on Earth*, Bloomsbury

Tarnopolski, Vladimir, 'Tabula Russia', 2016, https://tarnopolski.ru/en

Thoreau, Henry David, Journal, 12 October 1851, quoted in Todd Titon, Jeff, 'Thoreau's Ear', *Sound Studies*, 1(1), 1 February 2016, pp. 144–54, DOI: 10.1080/20551940.2015.1079973

Time and Tide, 'The Bell Design', https://timeandtidebell.org/the-bell-design/

Varda, Agnès, 2008, *The Beaches of Agnès*, Les Films du Losange

Vergette, Marcus, 2021, personal communication

Žalėnas, Gintautas, 'Cum Signo Campanae: The Origin of Bells in Europe and their Early Spread', Vytauto Didžiojo universiteto leidykla, *Sacrum et publicum*, 2013, pp. 67–94

Resonance (2)

Cappella Romana, 2019, 'Lost Voices of Hagia Sophia', cappellaromana.org

Cox, Trevor, 2014, *Sonic Wonderland: A Scientific Odyssey of Sound*, The Bodley Head

Jamie, Kathleen, 2003, 'Into the Dark', *London Review of Books*, 18 December 2003, https://www.lrb.co.uk/the-paper/v25/n24/kathleen-jamie/into-the-dark

Pentcheva, Bissera V., and Abel, Jonathan S., 'Icons of Sound: Auralizing the Lost Voice of Hagia Sophia', *Speculum*, October 2017

Pessoa, Fernando (Caeiro, Alberto), 2020, *The Complete Works of Alberto Caeiro*, W.W. Norton & Company

Rée, Jonathan, 1999, *I See a Voice: Deafness, Language and the Senses – A Philosophical History*, Metropolitan Books

Roomful of Teeth, 2017, *How a Rose*, https://roomfulofteeth.bandcamp.com/album/how-a-rose/. Listen also to a 2016 recording of *Es ist ein Ros entsprungen* by Voces8 and Jan Sandström, https://open.spotify.com/track/4nOyPDxuz3UYWRgAo5eon8?si=621a6a7e1a9d49b8

Service, Tom, 'The Timeless Power of Contemporary Choral Music', *The Listening Service*, BBC Radio 3, 17 October 2021, https://www.bbc.co.uk/sounds/play/m0010nx4

'World According to Sound': Sound Break: Hagia Sophia, 2020, https://soundcloud.com/worldaccordingtosound/114-sound-break-hagia-sophia

Frontiers

Bakker, Karen, 2022, *The Sounds of Life: How Digital Technology is Bringing Us Closer to the Worlds of Animals and Plants*, Princeton University Press

Battson, Ginny (undated), 'My World of Wordcraft – Neologisms for Rapidly Changing Times', Seasonlight (blog), https://seasonalight.com/for-the-love-of-planet-valens-my-neologisms/

Bratton, Benjamin, and Agüera y Arcas, Blaise, 'The Model is the Message', *Noema Magazine*, 12 July 2022, https://www.noemamag.com/the-model-is-the-message

Bruce, David, 2021, 'The DALL·E 2 of MUSIC?', https://www.youtube.com/watch?v=QNoDDD7B30U

De Abaitua, Matthew, 2022, 'The Dolittle Machine', BBC Radio 4, 30 May 2022, https://www.bbc.co.uk/programmes/m0017khn

Lomas, Tim (undated), 'The Lexicography', https://www.
drtimlomas.com/lexicography/cm4mi

Millman, Noah, 'A.I.s of Tumblr', *Gideon's Substack*, 15 June 2022,
https://gideons.substack.com/p/ais-of-tumblr

Monbiot, George, 2022, *Regenesis: Feeding the World Without
Devouring the Planet*, Penguin

Sacasas, L. M., 'LaMDA, Lemoine, and the Allures of Digital
Re-enchantment', *The Convivial Society* (substack), 28
June 2022, https://theconvivialsociety.substack.com/p/
lamda-lemoine-and-the-allures-of

Szenicer, Alexandre, et al., 'Seismic Savanna: Machine Learning
for Classifying Wildlife and Behaviours Using Ground-based
Vibration Field Recordings', *Remote Sensing in Ecology and
Conservation*, 9 November 2021

Silence

Barenboim, Daniel, 2006, 'In the Beginning was Sound', Reith
Lectures, Lecture 1, http://downloads.bbc.co.uk/rmhttp/
radio4/transcripts/20060407_reith.pdf

Beech, Hannah, 'Where Poets Are Being Killed and Jailed After
a Military Coup', *New York Times*, 25 May 2021, https://www.
nytimes.com/2021/05/25/world/asia/myanmar-poets.html

Berger, John, 1972, *Ways of Seeing*, BBC Two TV Series, and book,
Penguin

Breeden, Aurelien, 'From "Alive Among the Dead" to "Dead
Among the Living"', *New York Times*, 13 November 2021, https://
www.nytimes.com/2021/11/13/world/europe/france-2015-
attacks-trial-victims.html

Dasgupta, Partha, et al., 2021, *Final Report – The Economics of
Biodiversity: The Dasgupta Review* (independent report to the UK
government), https://www.gov.uk/government/publications/
final-report-the-economics-of-biodiversity-the-dasgupta-review

Goodman, Paul, 1972, *Speaking and Language*, excerpted
at themarginalian.org, https://www.themarginalian.
org/2015/01/13/paul-goodman-silence/

Griffiths, Jay, 1999, *Pip Pip: A Sideways Look at Time*, Flamingo

Hempton, Gordon, with Loften, Adam, and Vaughan-Lee, Emmanuel, 2018, 'Sanctuaries of Silence: An Immersive Listening Journey into the Hoh Rainforest', https://emergencemagazine.org/feature/sanctuaries-of-silence/

Heneghan, Liam, 'A Place of Silence', *Aeon*, 24 February 2020, https://aeon.co/essays/why-we-need-an-absence-of-noise-to-hear-anything-important

Hoban, Russell, 1980, *Riddley Walker*, Jonathan Cape

Imbler, Sabrina, 'The Maori Vision of Antarctica's Future', *New York Times*, 2 July 2021, https://www.nytimes.com/2021/07/02/science/antarctica-maori-exploration.html

Kagge, Erling, 2017, *Silence: In the Age of Noise*, Viking

Krznaric, Roman, 2020, *The Good Ancestor: How to Think Long Term in a Short-Term World*, W. H. Allen

Marías, Javier (trans. Margaret Jull Costa), 1996, *Tomorrow in the Battle Think on Me*, Harvill

Margalit, Avishai, 2002, *The Ethics of Memory*, Harvard University Press

Maysenhölder, W., et al., 2008, 'Sound Absorption of Snow', IBP Report 486, Fraunhofer Institut Bauphysik

Parshina-Kottas, Yuliya, et al., 'What the Tulsa Race Massacre Destroyed', *New York Times*, 24 May 2021, https://www.nytimes.com/interactive/2021/05/24/us/tulsa-race-massacre.html

Pomerantsev, Peter, 2014, *Nothing Is True and Everything Is Possible*, Public Affairs

Remnick, David, 'The Weakness of the Despot', interview with Stephen Kotkin, *The New Yorker*, 11 March 2022, https://www.newyorker.com/news/q-and-a/stephen-kotkin-putin-russia-ukraine-stalin

Riley, Charlotte L., 'The Empire Strikes Back', *New Humanist*, 10 December 2019

Sherrell, Daniel, 2021, *Warmth: Coming of Age at the End of Our World*, Penguin Random House

Smith, Clint, 2014, 'The Danger of Silence', TED Talk, https://www.ted.com/talks/clint_smith_the_danger_of_silence

Spooner-Lockyer, Kassandra, and Kilroy-Marac, Katie, 'Ten Things About Ghosts and Haunting', *Anthropology News*, 18

October 2021, https://www.anthropology-news.org/articles/
ten-things-about-ghosts-and-haunting/

Thoreau, Henry David, (1853) 2009, *The Journal 1837–1861*, New York
Review Books Classics

Tingley, Kim, 'Whisper of the Wild', *New York Times*, 18 March
2012, https://www.nytimes.com/2012/03/18/magazine/is-
silence-going-extinct.html

Twigger, Robert, 'Desert Silence', *Aeon*, 26 April 2013, https://aeon.
co/essays/how-the-sound-of-silence-rejuvenates-the-soul

Wickenden, Dorothy, 'Wendell Berry's Advice for a
Cataclysmic Age', *The New Yorker*, 28 February 2022,
https://www.newyorker.com/magazine/2022/02/28/
wendell-berrys-advice-for-a-cataclysmic-age

Weil, Simone (1950) 2009, *Waiting for God*, HarperCollins

Permissions

The author and the publisher have made every effort to trace copyright holders. Please contact the publisher if you are aware of any omissions

Bates, H.E., passage from *Through the Woods*. Reproduced with kind permission of Little Toller and the Estate of H.E. Bates.

Robert Haas version of Bashō. Reproduced in the UK with permission of Bloodaxe. Reproduced in US and Canada with permission of HarperCollins.

Michel Lara for his interpretation of Bashō.

Rumi, translated by Haleh Liza Gafori, *Gold*, New York Review Books, 2022. Reproduced with kind permission of Haleh Liza Gafori.

Index